Physics for Technology

with Applications in
Industrial Control Electronics

Daniel H. Nichols

DeVry Institute of Technology

Prentice
Hall

Upper Saddle River, New Jersey
Columbus, Ohio

Library of Congress Cataloging-in-Publication Data

Nichols, Daniel H.
    Physics for technology: with applications in industrial control electronics/Daniel H. Nichols
        p. cm.
    Includes index.
    ISBN 0-13-026602-7
    1. Physics I. Title

    QC23.2. N53 2002
    530′.024′62—dc21

                                                                    2001040921

Editor in Chief: Stephen Helba
Executive Editor: Frank I. Mortimer, Jr.
Media Development Editor: Michelle Churma
Production Editor: Louise N. Sette
Production Supervision: Clarinda Publication Services
Design Coordinator: Robin G. Chukes
Cover Designer: Thomas Borah
Cover photo: © FPG
Production Manager: Brian Fox
Marketing Manager: Tim Peyton

This book was set in Times Roman by The Clarinda Company. It was printed and bound by Courier Kendallville, Inc. The cover was printed by The Lehigh Press, Inc.

Pearson Education Ltd., *London*
Pearson Education Australia Pty. Limited, *Sydney*
Pearson Education Singapore Pte. Ltd.
Pearson Education North Asia Ltd., *Hong Kong*
Pearson Education Canada, Ltd., *Toronto*
Pearson Educación de Mexico, S.A. de C.V.
Pearson Education—Japan, *Tokyo*
Pearson Education Malaysia Pte. Ltd.
Pearson Education, *Upper Saddle River, New Jersey*

10 9 8 7 6 5 4 3 2 1
ISBN: 0-13-026602-7

DEDICATION

This book is dedicated to Devryn.

PREFACE

This book is intended as an introductory physics text for students majoring in engineering technology programs with particular emphasis in electronic applications. A background in algebra and AC/DC electronics is assumed. It was written in an effort to try to make clear the relevance of physics to a career in electronics. To achieve this, lengthy derivations and theoretical discussions on physics have been replaced with practical applications of physics in industry.

Unlike most texts in physics where a general concept is explained and leads to specific examples, I tried to take the opposite approach. Whenever possible, the discussion starts off with something the student is familiar with (for example, a speedometer, thermometer, etc.) and generalizes it to the physics behind these devices. This approach makes clear the motivation and relevance for studying physics. One of the unique features of this book is the incorporation of electronics into the general physics discussions. By doing so, this book serves as an introduction to the physics of sensors and an introduction to industrial control electronics. The text is intended for a one-semester course in physics using a judicious selection of chapters.

Acknowledgments

I would like to thank my colleagues at DeVry, Chicago: Dean Patrick O'Connor for editing the first draft and many helpful discussions, and Don Ingram for many of the photographs used throughout the book. I would also like to thank my good friend Penny for her help with the illustrations and editing the text. I would like to acknowledge the reviewers of this text: Steve Brown, Clackamas Community College (OR); Michael Crittenden, Genesse Community College (NY); Susan Ramlo, University of Akron (OH); and William Shi, DeVry Institute of Technology (NY). Lastly, I would like to thank my family for all their encouragement and support.

DANIEL H. NICHOLS

Contents

CHAPTER 6
Rotational Motion   91

CHAPTER 7
Machines   104

Fundamental Electronics Review

This chapter presents a review of fundamental electronics. It covers resistors, capacitors, and inductors, as well as some fundamental concepts to be used throughout the book. It is assumed that the reader already has a basic knowledge of AC/DC electronics; therefore, this chapter will not go into great depth and will simply serve as a review.

Figure 1.1

Electricity is the movement of charge through a conductor. Charges come in two "flavors": positive and negative. Electrons are negatively charged and orbit the atom. At the center of the atom are positively charged particles called *protons* and other particles with no charge, which are called *neutrons*. To make a charge move, place another charge by it. Like charges repel; unlike charges attract (Fig. 1.2). Two electrons placed near each other will try to move apart.

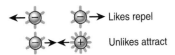

Figure 1.2
Like charges repel each other; unlike charges attract.

Charges never actually touch each other, but they experience each other through an electric field emanating from them. Electricity is by definition the movement of electrons through a conductor. To get electrons to flow, place a charge at the end of a conductor (Figs. 1.3 and 1.4).

Figure 1.3
Electricity is the movement of charge through a conductor.

Figure 1.4
To get electricity to flow, place a charge at the end of the conductor.

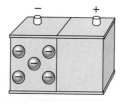

Figure 1.5
A battery is a storehouse of electrons.

A *battery* is a storehouse of electrons with one side full of electrons (the negative side) and the other side (the positive side) with fewer electrons (Fig. 1.5).

Connecting a wire to the negative side of the battery allows an escape route for the electrons to get away from each other. As the electrons escape, they begin to build up in the wire, and the flow from the battery will stop. (Fig. 1.6).

To continue the flow, a path to the positive side (where there are fewer electrons) needs to be made. Connecting a light bulb or anything else between the two battery terminals will complete the path back to the battery (Fig. 1.7).

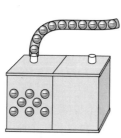

Figure 1.6
A wire connected to a battery will fill up with electrons, then the flow will stop.

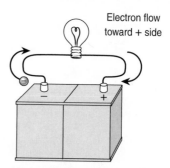

Electron flow toward + side

Figure 1.7
A complete path will allow the electrons to flow back to the battery.

The word *circuit* comes from the word *circle*. The electrons complete a path that returns them to their origin, similar to a circle.

Conductors *Conductors* are materials that allow electricity to flow easily. Materials such as copper and aluminum are conductors.

Insulators *Insulators* are materials that prevent electricity from flowing. Materials such as glass and rubber are insulators.

Semiconductors *Semiconductors* are materials whose conductivity falls somewhere between that of conductors and that of insulators. Different types of semiconductors can be put together in various combinations to either enhance or limit the flow of current, such as in the case of transistors or diodes.

OHM'S LAW

To get water to flow through a hose, we must apply pressure to it. To get electrons to flow through a wire, we must apply electrical pressure to them. Voltage can be thought of as electrical pressure. *Volt* (V) is the unit of voltage.

> Voltage = electrical pressure (volts)

Voltage and the electric field are related to each other. *Voltage* is a measure of the rate at which the electric field changes over some distance.

Water flowing through a pipe flows at some rate like gallons/minute. Electrons flow through a wire at a rate of (number of electrons)/second, called *amperes* or *amps* (A).

> Current = flow rate (amps)

The flow of water can be controlled using a faucet. The flow of electrons can be controlled using a resistor (Fig. 1.8). *Resistance* means opposition to flow and has a unit called *ohms* (Ω).

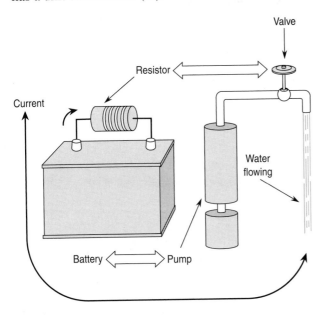

Figure 1.8
A plumbing analogy to electricity.

> Resistance = opposition to flow (ohms)

These three are related to each other by Ohm's law,

$$I = \frac{V}{R}$$

$$\text{Current} = \frac{\text{voltage}}{\text{resistance}}$$

This equation says that if the electrical pressure (voltage) is big and the opposition to the flow (resistance) is small, the flow (current) will be large.

EXAMPLE 1.1

A light bulb has a resistance of 2 Ω and is connected to a 12-V battery (Fig. 1.9). What is the current flowing through it?

$$V = 12\ V$$

$$R = 2\ \Omega$$

$$I = ?$$

$$I = \frac{V}{R}$$

$$I = \frac{12\ V}{2\ \Omega}$$

$$I = 6\ A$$

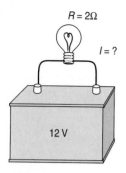

$R = 2\Omega$

$I = ?$

12 V

Figure 1.9

Figure 1.10
DC current flows in one direction.

AC/DC   *DC* stands for direct current. DC current moves in one direction (Fig. 1.10). A battery connected to a light bulb will produce a current flowing in one direction going from the negative to the positive terminal.

AC stands for alternating current. Alternating means that the current changes direction, that is, it moves back and forth (Fig. 1.11). The electricity in your house is AC current.

A moment later →

Figure 1.11
AC current changes direction.

AC current has some advantages over DC current because its voltage can be easily raised or lowered using a transformer (see below). The big gray cylindrical object on phone poles is a transformer. It lowers the voltage from a high voltage to 220 V and 110 V for your home (Fig. 1.12).

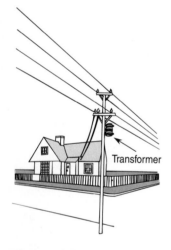

Transformer

Figure 1.12
AC power is transmitted to your home. The transformer lowers the voltage and delivers it to your house.

Resistors   *Resistors* reduce the flow of current in a circuit. They are analogous to faucets in plumbing because they reduce the flow of water (Fig. 1.13). A material inside a resistor causes the electrons to scatter when they move through it. The value of the resistance is determined by controlling the concentration of this scattering material in the resistor. Higher scattering means a higher resistance.

Some resistors can be used as sensors of temperature, force, humidity, magnetic fields, and so on because these quantities can affect the resistor's resistance. We shall see later in the book how to use resistors to measure these quantities as well as others.

Figure 1.13
A resistor scatters electrons flowing through it, causing a reduction in flow of the circuit.

Capacitors   *Capacitors* are devices that hold charge like a battery. Typically a capacitor consists of two separated metal plates (Figs. 1.14 through 1.17). If the capacitor is connected to a battery, charge runs out of the negative side of the bat-

Figure 1.14
Typically a capacitor consists of two separated metal plates.

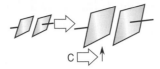

Figure 1.15
If the plates get bigger, the capacitance goes up.

Figure 1.16
If the separation between the plates gets smaller, the capacitance goes up.

tery and deposits itself on one plate of the capacitor. The other plate is connected to the positive side of the battery. Any free electrons on the positive plate will move into the positive side of the battery, to get as far away as possible from the electrons on the negative plate. If the capacitor is then disconnected from the battery, it will hold this charge on the negative plate. *Capacitance (C)* is a measure of how much charge the plates can hold.

The capacitance of a capacitor is described in the following equation:

$$C = \frac{\kappa \times A}{d} \quad \leftarrow \quad \textit{The unit of capacitance is the farad (F)}$$

$$\text{Capacitance} = \frac{\text{dielectric constant} \times \text{area of plates}}{\text{plate separation}}$$

Some capacitors may be used as sensors of proximity, distance, fluid level, and so on because these quantities can affect the capacitor's capacitance. We shall see later in the book how to use capacitors to measure these quantities.

Electromagnetic Induction   *Electromagnetic induction* is the generation of electricity with magnetism. When a magnetic field is changing in a coil of wire, a current will be induced in the wire (Fig. 1.18). This is the principle of an electrical generator.

Electromagnets   When a current is moving through a wire, a magnetic field is generated around it (Fig. 1.19).

If current flows through a coil, a magnetic field will be generated that looks very similar to the field of a bar magnet. This is an electromagnet (Figs. 1.20 and 1.21).

Inductors   *Inductors* are devices that oppose the current flow when the current is changing rapidly. An inductor is simply a coil of wire. When a current is changing rapidly through an inductor, the magnetic field generated by this current will change rapidly. By electromagnetic induction a countercurrent will be induced

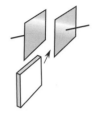

Figure 1.17
If something is inserted between the plates (a dielectric), the capacitance changes.

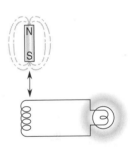

Figure 1.18
An electrical generator is made by changing a magnetic field through a coil. We can change the field through the coil by moving either the coil or the magnet.

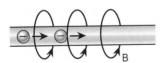

Figure 1.19
A magnetic field surrounds a current-carrying wire.

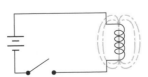

Figure 1.20
An electromagnet.

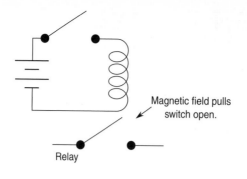

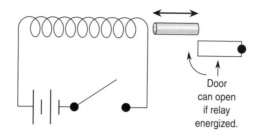

Magnetic field pulls
switch open.

Relay

Door
can open
if relay
energized.

Figure 1.21
An electromagnet can be used in many ways: as an electromechanical switch
(i.e., relay), as a door latch, or in a speaker, to drive the cone back and forth.

into the coil, which will oppose the incoming current. The inductor will therefore act like a resistor and reduce the current flow. Inductance is a measure of the inductor's ability to fight this changing current (Figs. 1.22–1.25).

The inductance of an inductor is given by the following formula:

$$L = \frac{\mu \times A \times N^2}{l} \quad \leftarrow \quad \textit{The unit of inductance is the henry (H)}$$

$$\text{Inductance} = \frac{\text{permeability} \times \text{area} \times \text{turns}^2}{\text{length}}$$

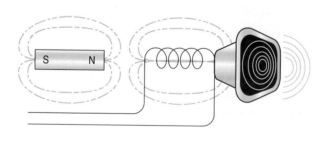

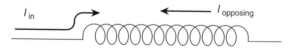

Figure 1.22
An inductor is a coil of wire that creates an opposing current when the incoming current is changing rapidly.

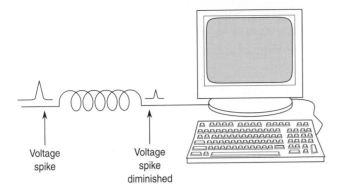

Voltage
spike

Voltage
spike
diminished

Figure 1.23
Inductors can be used as surge protectors, preventing voltage spikes from
entering your computer.

Figure 1.24
Adding a core to the
inductor will increase its
inductance.

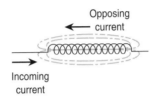

Opposing
current

Incoming
current

Figure 1.25
The inductor opposes a
rapidly-changing current
by using magnetic fields
generated in the coil to
produce a countercurrent
flow, which then opposes
the incoming current.

Anything that affects the permeability of an inductor will affect its inductance. Inductors can be used as sensors of proximity, distance, vibration, and so on because these quantities can affect inductance. Later in the book we shall see how to use inductors to measure these quantities.

Transformers A *transformer* is a set of coils coupled together through a magnetic field. A changing current passing through the primary coil generates a magnetic field, then passes through the secondary coil. Through electromagnetic induction, a current is generated in the secondary coil. The induced secondary voltage will be larger if there are more secondary coil turns than in the primary coil, in which case the transformer is a "step-up transformer." The induced secondary voltage will be smaller if there are fewer secondary coil turns than in the primary coil, in which case the transformer is a "step-down transformer" (Fig. 1.26).

Batteries A *battery* is a source of electrical pressure or voltage. The larger the voltage, the more current it can push through a circuit. How much charge a battery can store is determined by its amp-hour rating.

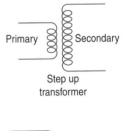

Primary   Secondary

Step up
transformer

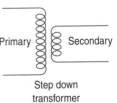

Primary   Secondary

Step down
transformer

Figure 1.26
A step-up and a step-
down transformer.

Amp-hours The *amp-hour* rating of a battery is the amount of charge a battery can store. The more charge the battery can store, the longer it can last when supplying a certain current. This number is normally printed on rechargeable batteries only.

$$Ah = amps \times hours$$

EXAMPLE 1.2

A battery has a 10-Ah rating. This means it can put out

10 A for 1 hour

5 A for 2 hours

2.5A for 4 hours

The more current used, the shorter the life of the battery (Fig. 1.27)

Battery	Voltage	Ampere-hour Rating	Mass, g
Primary			
"pen-light"	1.5	0.58	20
D-cell	1.5	3.0	90
#6 dry cell	1.5	30	860
#2U6 battery	9.0	0.325	30
#V-60 battery	90	0.47	580
#1 mercury cell	1.4	1.0	12
#42 mercury cell	1.4	14	166
Secondary (rechargeable)			
#CD-21 nickel-cadmium	6.0	0.15	52
#CD-29 nickel-cadmium	12.0	0.45	340
lead-acid car battery	6.0	84	13 000
lead-acid car battery	12.0	96	25 000

Figure 1.27
Amp-hour ratings for some typical batteries.

Power supplies A power supply is like a battery that never runs out of power because it is plugged into the wall. Some power supplies are adjustable in voltage, and for all of them there is a limit to how much current they can put out at a given voltage. This is called *compliance*.

Compliance *Compliance* is the maximum current a supply can put out before the output voltage begins to drop.

EXAMPLE 1.3

A 5-V power supply has a compliance of 1.5 A. If a circuit draws more than 1.5 A from the supply, the voltage will drop below 5 V, possibly damaging the supply.

Schematic Symbols   Schematic symbols are shorthand symbols that represent components in electronics. They save time and space when one is drawing a circuit. Some of the symbols are given in Figure 1.28.

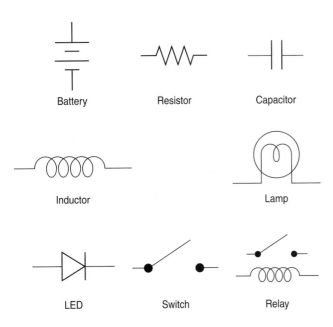

Figure 1.28
Schematic symbols.

 ## MEASURING VOLTAGE, CURRENT, AND RESISTANCE

A *multimeter* is an instrument designed to measure voltage, current, and resistance. The three measurements are illustrated in Figures 1.29 through 1.31.

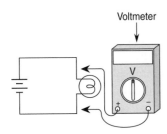

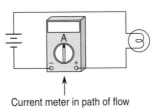

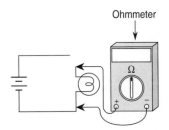

Figure 1.29
Measuring voltage. To measure voltage, the voltmeter is placed across the component whose voltage is being measured.

Figure 1.30
Measuring current. To measure current, the current meter is placed in the same path as the component with which the current is being measured.

Figure 1.31
Measuring resistance. To measure resistance, the ohm meter is placed across the component with the component taken out of the circuit.

Power   *Electrical power* is the rate at which electrical energy is being used. It is measured in watts (W).

$$P = I \times V$$

$$\text{Power} = \text{current} \times \text{volts}$$

EXAMPLE 1.4

A 9-V battery is connected to a radio that draws 0.1 A (Fig. 1.32). What is the power consumed by this radio?

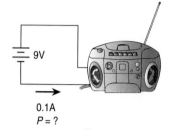

9V

0.1A
P = ?

Figure 1.32

$$V = 9 \text{ V}$$

$$I = 0.1 \text{ A}$$

$$P = ?$$

$$P = 1 \times V$$

$$P = 0.1 \text{ A} \times 9 \text{ V}$$

$$P = 0.9 \text{ W}$$

Electrical Energy   The power company charges you by how much electrical energy you use. The power meter on your house records the amount of energy you use by counting the number of kilowatt-hours (kWh) consumed. Kilowatt-hours indicate the number of kilowatts used and how long they were used:

$$\text{kWh} = \text{kW} \times \text{h}$$

EXAMPLE 1.5

Suppose you blow-dry your hair with a hair dryer that is rated at 1.2 kW. If you operate the dryer for half an hour, the number of kilowatts used is:

$$\text{kWh} = (1.2 \text{ kW})(1/2 \text{ h})$$

$$\text{kWh} = 0.6 \text{ kWh}$$

The power company charges a certain rate, $\dfrac{\text{cents}}{\text{kWh}}$, for energy consumed. To determine the cost for the use of so many kilowatt-hours consumed, apply the following equation:

$$\text{Cost} = \text{rate} \times \text{kWh}$$

EXAMPLE 1.6

On average the power companies charge 9 cents/kWh. Suppose you leave a 100-W lightbulb on overnight for 8 hours. How much will this cost you?

$$\text{rate} = 9 \text{ cents/kWh}$$

$$t = 8 \text{ h}$$

$$P = 100 \text{ W} = 0.1 \text{ kW}$$

$$\text{cost} = \text{kWh} \times \text{rate}$$

$$\text{cost} = 0.1 \text{ k\cancel{W}} \times 8\cancel{h} \times 9 \frac{\text{cents}}{\text{k\cancel{W} \cancel{h}}}$$

$$\text{cost} = 7.2 \text{ cents}$$

 ## PROBLEM-SOLVING TECHNIQUES

A lot of the difficulty in solving word problems occurs in trying to translate the English into the mathematics. Listed below is a set of steps to help in the solution of these problems.

1) As you are reading through the problem, write down the values of the quantities given.

For example: A 50-W light bulb is connected across 110 V. How much current does it draw?

As you are reading "50-W," write down P = 50 W. When you read "across 110 V," write down V = 110 V. When you read "How much current does it draw?" write down I = ?

2) Make a sketch of the problem, labeling everything to help you visualize what is being asked (Fig. 1.33).

3) Work the problem in a vertical column, organizing the information and writing down everything that is relevant. List all the relevant equations (Fig. 1.34).

4) Work out any algebra first before plugging in numbers.

$$P = I \times V$$

$$I = ?$$

$$\frac{P}{V} = \frac{I \times \cancel{V}}{\cancel{V}}$$

$$I = \frac{P}{V}$$

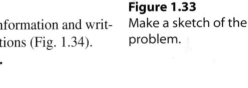

Figure 1.33
Make a sketch of the problem.

$V = 110 \text{ V}$
$P = 50 \text{ W}$
$I = ?$
$P = I \times W$

Figure 1.34
Write down everything relevant.

5) Make sure the quantities have the correct units when plugging into formulas.

$P = 75 \text{ W}$ $\leftarrow$ *P* is in watts, not in any other unit such as kilowatts or milliwatts.

$V = 110 \text{ V}$ $\leftarrow$ *V* is in volts.

$$I = \frac{P}{V}$$

$$I = \frac{75 \text{ W}}{110 \text{ V}}$$

$$I = 0.68 \text{ A}$$

6) Check to see if the answer makes sense both numerically and unit-wise.

If a question asks for current, the answer must have units of amps. If you got a different unit, you probably did something wrong.

EXAMPLE 1.7

An elevator motor runs on 220 V, draws 25 A of current, and operates 2.5 hours every day. What is the power drawn by this motor, and how much does it cost to operate it per day if the rate is 9 cents/kWh? Figure 1.35 shows the right and wrong ways to solve the problem.

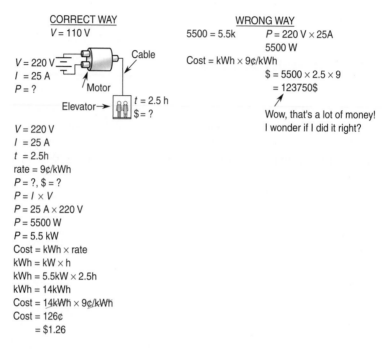

CORRECT WAY
$V = 110$ V

$V = 220$ V
$I = 25$ A
$P = ?$

Motor

Cable

Elevator

$t = 2.5$ h
$\$ = ?$

$V = 220$ V
$I = 25$ A
$t = 2.5$h
rate = 9¢/kWh
$P = ?, \$ = ?$
$P = I \times V$
$P = 25$ A $\times$ 220 V
$P = 5500$ W
$P = 5.5$ kW
Cost = kWh $\times$ rate
kWh = kW $\times$ h
kWh = 5.5kW $\times$ 2.5h
kWh = 14kWh
Cost = 14kWh $\times$ 9¢/kWh
Cost = 126¢
= $1.26

WRONG WAY
5500 = 5.5k $P = 220$ V $\times$ 25A
                 5500 W
Cost = kWh $\times$ 9¢/kWh
         $\$ = 5500 \times 2.5 \times 9$
         = 123750$

Wow, that's a lot of money!
I wonder if I did it right?

Figure 1.35
The right and wrong way to approach the problem.

CHAPTER SUMMARY

$$I = \frac{V}{R}$$

Current = $\dfrac{\text{voltage}}{\text{resistance}}$

$$C = \frac{\kappa \times A}{d}$$ ← The unit of capacitance is the farad (F)

Capacitance =
$$\frac{\text{dielectric constant} \times \text{area of plates}}{\text{plate separation}}$$

$$L = \frac{\mu \times A \times N^2}{l}$$ ← The unit of inductance is the henry (H)

inductance = $\dfrac{\text{permeability} \times \text{area} \times \text{turns}^2}{\text{length}}$

Ah = amps $\times$ hours

$$P = I \times V$$

Power = current $\times$ volts

kWh = kW $\times$ h

Kilowatt-hours, or kWh, is the number of kilowatts used and how long they were used.

$$\text{Cost} = \text{rate} \times \text{kWh}$$

Problem-Solving Tips:

■ When solving problems dealing with Ohm's law, do not calculate in kilo-ohms or millivolts or mil-liamperes; convert all values to their base units—ohms, volts, or amps—before plugging into Ohm's law.

■ Try to follow the six-step approach outline above when solving the problems.

PROBLEMS

1. What makes current flow in a circuit?

2. If a flashlight bulb has a resistance of 30 Ω and is connected to two D cells, creating a voltage of 3 V, what is the current flowing through the bulb?

3. What is the power consumed in problem 2?

4. If a car battery has an amp-hour rating of 96 Ah, and each time it is started it draws 250 A for a period of 5 seconds, how many times can it start the car without being recharged? Assume the charging system on the car is not working.

5. Go around your house and determine the total kilowatt-hours of energy consumed over a period of a few hours. Compare this with the meter reading on the outside of your house. Does it agree?

6. Suppose you wish to change the brightness of a light. How could you do this using a resistor?

7. Design a test system to measure the power consumed by a motor driving an elevator.

8. Suppose your car battery is dead and the battery has an amp-hour rating of 100 Ah. How long will it take to fully charge this dead battery if the charger is set to a charge rate of 5 A?

9. If something moves between the plates of a capacitor while the capacitance is being measured, what will happen to the capacitance?

10. If a metal object moves into the core of an air core inductor, what will happen to the inductance?

11. If the power company charges 10 cents/kWh, how much does it cost you over a year's time to blow-dry your hair for 5 minutes a day with a 1-kW hair dryer? Be careful of the units!

12. Suppose a window alarm is designed with an inductor and a movable metal core. Draw a sketch of how this might work.

13. Design a proximity sensor using a capacitor to determine if a refrigerator door is open.

14. How long can an electric car travel if the 12-V motor draws 30 A, and the batteries contain 500 Ah of charge?

15. How far does the car in problem 14 travel if it is moving at 55 miles per hour? Hint: look at the units.

Units

Units are based on standards wherein we define measurable quantities, such as length, mass, time, and so on. The units vary depending on the size of things we are measuring. For example, we would not measure the distance across the United States in inches; that is much too small a unit. Instead we would use a unit such as miles or kilometers. It is very handy to be able to convert back and forth between like units, such as miles, inches, and kilometers. This chapter will present some of the various ways of measuring, recording, and converting between units.

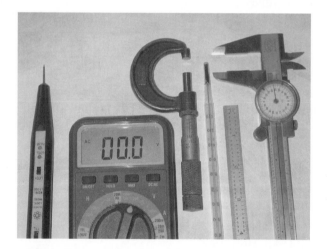

Figure 2.1

SCIENTIFIC NOTATION

Many measurements are written in *scientific notation,* which is a method of writing big and small numbers in a compact way. In this notation, all numbers are written in a power of ten.

$$10 = 1.0 \times 10^1$$
$$100 = 1.0 \times 10^2$$
$$1,000 = 1.0 \times 10^3$$

The number above the 10 is called an *exponent,* which tells how many zeros come after the 1.

For numbers such as 1/10 or 1/100, we use a minus sign in the exponent.

$$\frac{1}{10} = 1.0 \times 10^{-1}$$

$$\frac{1}{100} = 1.0 \times 10^{-2}$$

$$\frac{1}{1,000} = 1.0 \times 10^{-3}$$

The number in the exponent now tells how many zeros follow the 1 in the bottom of the fraction.

Numbers such as 638,000 or 0.00012 can be written in scientific notation.

$638,000 = 6.38 \times 10^{5}$ ← Decimal moved to the left five places

$0.00012 = 1.2 \times 10^{-4}$ ← Decimal moved to the right four places

Place the decimal after the first nonzero digit on the left and count the number of places the decimal was moved over. Use that number in the exponent. If the decimal is moved to the left, the exponent is positive. If the decimal is moved to the right, the exponent is negative.

UNITS OF LENGTH

In the English system, length units are miles, yards, feet, and inches. Another system is the SI system. SI stands for the Système International (International System, in English). The metric system is part of the SI system. The basic unit of length in the SI is the meter. All units of length are based on it.

In the early part of the twentieth century, the standard of measuring length was defined using a metal bar that was kept in a vault somewhere in France. It became necessary to devise a generic method of measuring length so the whole world could have access to it. To alleviate this problem, the meter is defined as the distance light travels in a vacuum during the time interval of $\frac{1}{299,792,458}$ second. This may seem like a strange way to define a meter, but it eliminates the need to have a single object stored in a vault somewhere, which acts as the definition of what "1 meter long" actually is.

In the metric system, length units are kilometers, meters, centimeters, and millimeters. The metric system is much easier to use than the English system of units because everything is grouped into powers of ten (Fig. 2.2).

- The mks system is part of the SI system. All units are recorded and calculated in meters, kilograms, and seconds (mks).
- In the English system, calculations are done using pounds, slugs, feet, and seconds. This system is sometimes called the fps system, or the foot-pound-second system.

In the metric system, all units are grouped into powers of ten using prefixes.

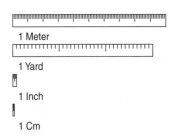

1 Meter

1 Yard

1 Inch

1 Cm

Figure 2.2
Comparison of various length units.

Table 2.1
Table of Metric Prefixes

$$1 \text{ giga} = 1 \text{ G} = 1 \times 10^9$$
$$1 \text{ mega} = 1 \text{ M} = 1 \times 10^6$$
$$1 \text{ kilo} = 1 \text{ k} = 1 \times 10^3$$
$$\text{Unity} = 1 = 1 \times 10^0$$
$$1 \text{ milli} = 1 \text{ m} = 1 \times 10^{-3}$$
$$1 \text{ micro} = 1 \mu = 1 \times 10^{-6}$$
$$1 \text{ nano} = 1 \text{ n} = 1 \times 10^{-9}$$
$$1 \text{ pico} = 1 \text{ p} = 1 \times 10^{-12}$$

1 km = 1000 m	1 km is approximately 0.6 miles
1 m = 100 cm	1 cm is approximately 0.4 inches
$1 \text{ mm} = \dfrac{1}{1,000} \text{ m}$	1 mm is approximately the thickness of a dime
$1 \ \mu\text{m} = \dfrac{1}{1,000,000} \text{ m}$	$1 \ \mu\text{m}$ is approximately $\dfrac{1}{100}$ the width of a human hair

Mechanical Measuring Instruments

The choice of the measuring instrument depends on the accuracy needed. A standard machinist's rule can measure accurately to the smallest division on the scale, typically 0.01 in or 0.1 mm (Fig. 2.3).

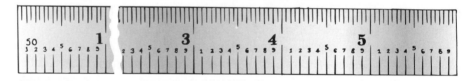

Figure 2.3
A machinist's scale is accurate to 0.01 in.

For better accuracy, a vernier caliper can measure both inside and outside diameters accurately to 0.001 in (Fig. 2.4).

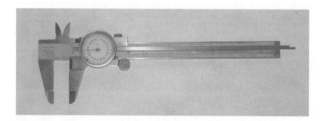

Figure 2.4
A vernier caliper measures accurately to 0.001 in.

A high-precision micrometer can measure accurately to 0.0001 in., or 0.002 mm (Fig. 2.5).

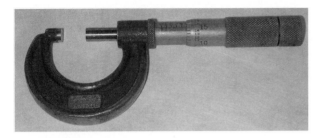

Figure 2.5
A high-precision micrometer measures accurately to 0.0001 in.

Recording and Calculating with Measurements

When we record a measurement, the number we write down should convey how accurately the measurement was made. For example, 2.1 in and 2.10 in mean the same thing mathematically but not in terms of measurements. The number 2.10 in means the number was measured to one-hundredth of an inch. The number 2.1 in means the number was measured to one-tenth of an inch.

Suppose a measurement is made with a voltmeter that displays the measurement as 2.1 V. If the real value of the measurement is 2.15 V or greater, the meter would read 2.2 V. If the value is less than 2.05 V, the meter would read 2.0 V. Therefore, 2.1 V means the measurement was made with an uncertainty of ± 0.05 V, and 2.10 V means the measurement was made with an uncertainty of ± 0.005 V.

$$\text{The uncertainty is } \pm \frac{1}{2} \text{ the smallest scale division.}$$

Significant Digits

A *significant digit* is a digit in a number that has been measured. When calculating with numbers, we must take into account the accuracy of the numbers in the calculation. The following rules have been adopted to convey how accurately something has been measured:

1. All nonzero numbers are significant.
2. Zeros are significant if . . .
   a. They have a bar placed over them.
   b. They lie between significant digits.
   c. They follow the decimal and a significant digit.

Number	Significant digits
2.1 in	2
2.10	3
0.01 in	1
350	2
2.01 × 104	3
3$\overline{0}$0	2
40$\overline{0}$	3

Rounding Off

When rounding off a number, if the first digit dropped is less than 5, round down; if it is 5 or greater, round up.

EXAMPLE 2.1

Round off 254.735 to the nearest 100, 10, 1, 0.1, and 0.01.

254.735 = 300	Nearest 100
254.735 = 250	Nearest 10
254.735 = 255	Nearest 1
254.735 = 254.7	Nearest 0.1
254.735 = 254.74	Nearest 0.01

Addition/Subtraction

When quantities are added or subtracted, first round the numbers to the correct number of significant digits, then round the result to the decimal place of the number with the fewest decimal places in the addition or subtraction.

$$2 \text{ in} + 1.4 \text{ in} = 3 \text{ in}$$
$$2.0 \text{ in} + 1.4 \text{ in} = 3.4 \text{ in}$$
$$2 \text{ in} - 1.4 \text{ in} = 1 \text{ in}$$

Multiplication/Division

When quantities are multiplied or divided, round the result so that it has no more significant digits than the number with the least significant digits in the multiplication or division.

$$(28.0 \text{ ft})(6 \text{ ft}) = 200 \text{ ft}^2$$

$$(28.0 \text{ ft})(6.0 \text{ ft}) = 170 \text{ ft}^2$$

$$\frac{28.0 \text{ ft}}{6 \text{ ft}} = 5$$

$$\frac{28.0 \text{ ft}}{6.0 \text{ ft}} = 4.7$$

An exception to this is the case when conversions are used. For example, 1 ft = 12 in. is exact and should be treated as exact, that is, having an infinite number of significant digits.

$$0.2563 \text{ ft}^2 \left(\frac{12 \text{ in}}{1 \text{ ft}} \right)^2$$

$$= 0.2563 \text{ ft}^2 \left(\frac{144 \text{ in}^2}{1 \text{ ft}^2} \right)$$

$$= 36.91 \text{ in}^2$$

Calibration

Calibration means to compare a measurement device against a known standard. Calibrating a measurement device means to adjust its output so that it displays correctly what it is measuring. A correctly calibrated instrument is very important, especially in the medical field, where an incorrectly calibrated instrument could cost someone's life.

Unit Conversions

To convert between the two systems, use the following conversions:

1 in = 2.54 cm

1 mi = 100 cm

1 cm = 10 mm

1 mi = 5,280 ft

1 yd = 3 ft

1 ft = 12 in

Suppose you are given a measurement in feet and want to know how many inches that equals. Use the bracket method.

The Bracket Method

The bracket method is a way to convert between any type of like units.
Convert 4.20 ft into inches.
To use the bracket method, apply the following steps:

Step 1   Write down the conversion.

$$1 \text{ ft} = 12 \text{ in}$$

Step 2   Write down what is to be converted with a bracket next to it.

$$4.20 \text{ ft} \, (-)$$

Step 3   Fill in the bracket with the conversion, placing the unit that is on the outside of the bracket in the bottom of the bracket and the unit it is equal to in the top part of the bracket.

$$4.20 \text{ ft} \left(\frac{12 \text{ in}}{1 \text{ ft}} \right) = 50.4 \text{ in}$$

EXAMPLE 2.2

Convert 30.0 in. into feet.

$$1 \text{ ft} = 12 \text{ in}$$

$$30.0 \text{ in} \left(\frac{1 \text{ ft}}{12 \text{ in}} \right) = 2.50 \text{ ft}$$

Notice that the unit on the outside of the bracket is in inches; therefore, the unit in the bottom of the bracket must also be in inches in order for inches to cancel out, leaving feet. It is not necessary to try to remember what goes in the top or bottom; just remember to place the unit that is on the outside of the bracket in the bottom of the bracket. Or if the unit on the outside is in the bottom, place the same unit inside the bracket in the top.

EXAMPLE 2.3

Convert 25 cm into meters.

$$1 \text{ m} = 100 \text{ cm}$$

$$25 \text{ cm} \left(\frac{1 \text{ m}}{100 \text{ cm}} \right) = 0.25 \text{ m}$$

EXAMPLE 2.4

Convert 25 miles per hour into miles per minute.

$$1 \text{ h} = 60 \text{ min}$$

$$25 \frac{\text{mi}}{\text{h}} \left(\frac{1 \text{ h}}{60 \text{ min}} \right) = 0.42 \frac{\text{mi}}{\text{min}}$$

The bracket method can be used when there is no direct conversion from one unit to another. To perform such a conversion, use the bracket method twice.

Multiple Conversions

Multiple conversions are used when no direct conversion exists. For example, convert 32.2 yd into inches. We know that:

$$1 \text{ yd} = 3 \text{ ft}$$
$$1 \text{ ft} = 12 \text{ in}$$

We do not have a direct conversion from yards to inches. We can, however, convert yards to feet, then feet into inches.

$$32.2 \text{ yd} \left(\frac{3 \text{ ft}}{1 \text{ yd}} \right) \left(\frac{12 \text{ in}}{1 \text{ ft}} \right) = 1,160 \text{ in}$$

In the first bracket, yards canceled out, leaving feet. Therefore, feet is on the out-side of the second bracket, and we should put feet in the bottom of the second bracket.

EXAMPLE 2.5

Convert 2.00 m into inches using the following conversions:

$$1 \text{ m} = 100 \text{ cm}$$
$$1 \text{ in} = 2.54 \text{ cm}$$

There is no direct conversion from meters to inches, so we will have to perform a double conversion.

$$2.00 \text{ m} \left(\frac{100 \text{ cm}}{1 \text{ m}} \right) \left(\frac{1 \text{ in}}{2.54 \text{ cm}} \right)$$
$$= 78.7 \text{ in}$$

Measuring Length Electronically

There are many ways to measure length or distance. Your accuracy requirements determine which type of instrument to use.

Electronic Micrometer An electronic micrometer is accurate to 0.0001 in (Figs. 2.6 and 2.7).

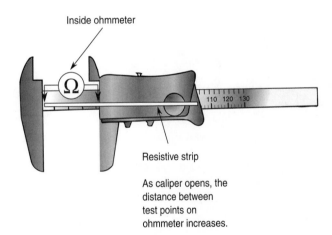

Inside ohmmeter

Resistive strip

As caliper opens, the distance between test points on ohmmeter increases.

Figure 2.6
An electronic micrometer.

Figure 2.7
Electronic micrometers and calipers contain a resistive strip. As the barrel or caliper is moved back, the distance between measuring points on the resistive strip increases, and therefore the resistance measured will increase. This resistance is converted into a distance and displayed.

Ultrasonic Tape Measure An ultrasonic tape measure, which is accurate to a few tenths of an inch over a distance of tens of feet, can be bought in any hardware store (Figs. 2.8 and 2.9).

Figure 2.8
Ultrasonic tape measure (courtesy of Zircon).

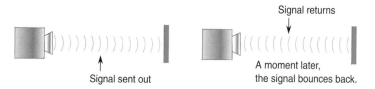

Figure 2.9
The ultrasonic tape measure works by sending a sound wave and timing how long it takes to return after bouncing off an object whose distance is being measured. The longer it takes, the farther away the object.

Laser Ranging Surveyors commonly use laser ranging (Figs. 2.10 and 2.11).

Figure 2.10
A commercial laser ranger (courtesy of Zircon).

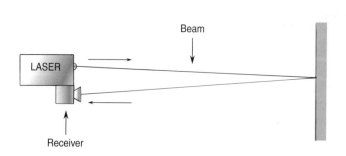

Figure 2.11
Laser ranging works by sending a light beam and timing how long it takes to return after bouncing off the object being measured. The longer it takes, the farther away the object. Background light does not interfere with the measurement because the detector is sensitive only to the frequency of the laser.

Capacitive Sensors   As the spacing between plates on a capacitor decreases, the capacitance increases. This change in capacitance can be used to determine the relative change in plate separation. By anchoring one of the plates to a fixed object and the other to an object in motion, the object in motion can be monitored by measuring the capacitance (Fig. 2.12).

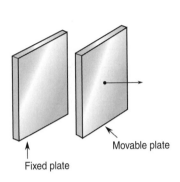

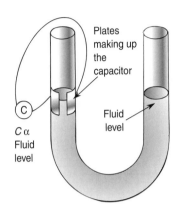

Figure 2.12
A variable capacitor is used to measure displacement.

Figure 2.13
A capacitor is used to measure the level of a fluid.

A capacitor can also measure displacement using a change in the dielectric's position between the two plates. As the dielectric is inserted into the capacitor, its capacitance will rise. This method is sometimes used to measure the height of a fluid in a tube (Fig. 2.13).

Inductive Sensors   In a variable inductor, as the core is slid in and out the inductance will change. By attaching the coil to a fixed object and the core to a movable object, the motion of the latter can be determined by measuring the change in inductance (Fig. 2.14).

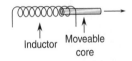

Figure 2.14
A variable inductor is used to measure length. As the core is moved out, the inductance will decrease.

 SURFACE AREA

If you have ever looked at heat sinks mounted on IC chips, the fins are designed to increase the surface area of the chip. A large surface area will diffuse heat faster than a small one.

Area has units of length^2 (spoken as "square meters" or "square feet," etc.).

$$1 \text{ m}^2 = 10,000 \text{ cm}^2$$
$$1 \text{ ft}^2 = 144 \text{ in}^2$$
$$1 \text{ cm}^2 = 0.155 \text{ in}^2$$
$$1 \text{ ft}^2 = 929 \text{ cm}^2$$

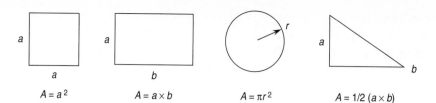

Figure 2.15
Areas of various shapes.

EXAMPLE 2.6

What is the area of a square with sides that are of 2.0 in in length?

$$A = a^2$$
$$A = (2.0 \text{ in})^2 = 4.0 \text{ in}^2$$

EXAMPLE 2.7

What is the area of a rectangle with a height of 2.5 in and a width of 3.0 in?

$$A = a \times b$$
$$A = (2.5 \text{ in}) (3.0 \text{ in})$$
$$A = 7.5 \text{ in}^2$$

EXAMPLE 2.8

What is the area of a circle with a radius of 3.00 cm?

$$A = \pi r^2$$
$$A = \pi (3.00 \text{ cm})^2$$
$$A = 28.3 \text{ cm}^2$$

EXAMPLE 2.9

What is the area of a right triangle with a height of 11 cm and a base of 12 cm?

$$A = \frac{1}{2} ab$$

$$A = \frac{1}{2} (11 \text{ cm}) (12 \text{ cm})$$

$$A = 66 \text{ cm}^2$$

Surface Area of Three-Dimensional Objects

The surface area of a three-dimensional shape is the sum of the areas of each of its sides.

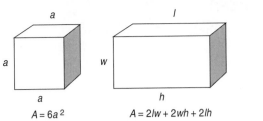

$A = 6a^2$ $A = 2lw + 2wh + 2lh$ $A = 2\pi r^2 + 2\pi rh$ $A = 4\pi r^2$

Figure 2.16
Surface area of various shapes.

EXAMPLE 2.10

What is the surface area of a cube with sides that are 2.0 cm long?

$$A = 6a^2$$
$$A = 6\,(2.0\text{ cm})^2$$
$$A = 24\text{ cm}^2$$

EXAMPLE 2.11

What is the surface area of a rectangular prism with a length of 2.5 in, a width of 3.5 in, and a height of 5.1 in?

$$A = 2lw + 2wh + 2lh$$
$$A = 2(2.5\text{ in})\,(3.5\text{ in}) + 2(3.5\text{ in})\,(5.1\text{ in}) + 2(2.5\text{ in})\,(5.1\text{ in})$$
$$A = 79\text{ in}^2$$

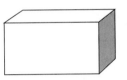

EXAMPLE 2.12

What is the surface area of a sphere with a radius of 3.00 cm?

$$A = 4\pi r^2$$
$$A = 4\pi(3.00\text{ cm})^2$$
$$A = 113\text{ cm}^2$$

EXAMPLE 2.13

What is the surface area of a Pringles® potato chip can with a radius of 3.00 cm and a height of 9.00 cm, excluding the ends?

$$V = \pi r^2 h$$
$$V = \pi(3.00\text{ cm})^2\,(9.00\text{ cm})$$
$$V = 254\text{ cm}^3$$

The Bracket Method Revisited

When converting units that involve powers, the bracket must be raised to the same power as the unit being converted. In the case of area conversions, the bracket must be squared.

> Remember, when raising a bracket to a power, everything in the bracket is raised to that power.

EXAMPLE 2.14

Convert 3 m² to square centimeters.

$$3 \text{ m}^2 \left(\frac{100 \text{ cm}}{1 \text{ m}} \right)^2$$

$$= 3 \text{ m}^2 \frac{100^2 \text{ cm}^2}{\text{m}^2}$$

$$= 30,000 \text{ cm}^2$$

EXAMPLE 2.15

Convert 12.3 cm² to square meters.

$$12.3 \text{ cm}^2 \left(\frac{1 \text{ m}}{100 \text{ cm}} \right)^2$$

$$= 12.3 \text{ cm}^2 \frac{1 \text{ m}^2}{100^2 \text{ cm}^2}$$

$$= 1.23 \times 10^{-3} \text{ m}^2$$

EXAMPLE 2.16

Convert 13.2 ft² to square inches.

$$13.2 \text{ ft}^2 \left(\frac{12 \text{ in}}{1 \text{ ft}} \right)^2$$

$$= 13.2 \text{ ft}^2 \frac{12^2 \text{ in}^2}{\text{ft}^2}$$

$$= 1,9\overline{0}0 \text{ in}^2$$

VOLUME CALCULATIONS

A volume calculation involves a length times length times length, or length³, such as the volume of a box where volume is determined by width times length times height.

Volume Units

SI

$$1 \text{ liter (L)} = 1{,}000 \text{ cm}^3$$
$$1 \text{ m}^3 = 10^3 \text{ liters}$$
$$1 \text{ cm}^3 = 1{,}000 \text{ mm}^3$$

English

$$1 \text{ gal} = 4 \text{ qt}$$
$$1 \text{ gal} = 8 \text{ pints}$$
$$1 \text{ gal} = 128 \text{ fl. oz.}$$
$$1 \text{ ft}^3 = 7.48 \text{ gal}$$

English to SI

$$1 \text{ ft}^3 = 28.3 \text{ liters}$$
$$25.3 \text{ ft}^3 = 1 \text{ m}^3$$
$$1 \text{ gal} = 3.785 \text{ liters}$$

Figure 2.17
A comparison of various units.

  $V = a^3$  $V = l \times w \times h$ $V = \pi r^2 h$ $V = \frac{4}{3}\pi r^3$

Figure 2.18
Volume of various shapes.

EXAMPLE 2.17

Find the radius of a baseball with a volume of 33.5 in³.

$$V = \frac{4}{3}\pi r^3$$

$$\frac{3}{4\pi} V = \frac{\cancel{3}}{\cancel{4\pi}} \frac{\cancel{4}}{\cancel{3}} \cancel{\pi} r^3$$

$$r^3 = \frac{3}{4\pi} V$$

$$r = \left(\frac{3}{4\pi} V\right)^{\frac{1}{3}}$$

$$= \left(\frac{3}{4\pi} 33.5 \text{ in}^3\right)^{\frac{1}{3}}$$

$$= 2.00 \text{ in}$$

Volume Unit Conversions—The Bracket Method

When converting units that involve powers, the bracket must be raised to the same power as the unit being converted. In the case of volume conversions, the bracket must be cubed.

> Remember, when raising a bracket to a power, everything in the bracket is raised to that power.

EXAMPLE 2.18

Convert 25 in^3 into cubic feet.

$$1 \text{ ft} = 12 \text{ in}$$

$$25 \text{ in}^3 \left(\frac{1 \text{ ft}}{12 \text{ in}} \right)^3$$

$$= 25 \text{ in}^3 \frac{\text{ft}^3}{12^3 \text{ in}^3}$$

$$= 0.014 \text{ ft}^3$$

EXAMPLE 2.19

Convert 1.2 m^3 into cubic centimeters.

$$1 \text{ m} = 100 \text{ cm}$$

$$1.2 \text{ m}^3 \left(\frac{100 \text{ cm}}{1 \text{ m}} \right)^3$$

$$= 1.2 \text{ m}^3 \frac{100^3 \text{ cm}^3}{\text{m}^3}$$

$$= 1.2 \times 10^6 \text{ cm}^3$$

EXAMPLE 2.20

Convert 2.5 in^3 into cubic feet.

$$1 \text{ ft} = 12 \text{ in}$$

$$2.50 \text{ in}^3 \left(\frac{1 \text{ ft}}{12 \text{ in}} \right)^3$$

$$= 2.50 \text{ in}^3 \frac{\text{ft}^3}{12^3 \text{ in}^3}$$

$$= 1.4 \times 10^{-3} \text{ ft}^3$$

Measuring Volume

One way to measure the volume of an object is to submerge it in water and measure how much water is displaced (Fig. 2.19).

MASS AND WEIGHT

Mass and weight are sometimes referred to as the same thing. This is not quite correct, however, as will be explained in more detail in a later chapter. For now, we will use simple definitions of mass and weight.

Mass   The *mass* of an object is the measure of the amount of matter contained in it. Its basic unit of measurement is the gram (g). A small paper clip has the mass of about 1 g.

Weight   The *weight* of an object refers to how heavy it is or how strongly it is attracted to the earth because of gravity.

Density   *Density* is a measure of how much mass or weight there is in a given volume. For example, a little bit of lead weighs a lot. It has a high density.

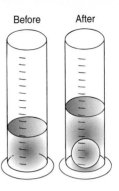

Figure 2.19
To measure the volume of an object, place the object in a full container of water and measure the amount of water displaced. The volume of the object equals the volume of water displaced.

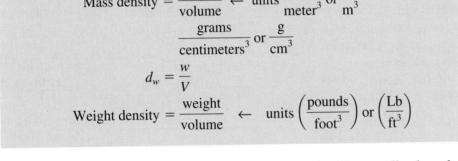

$$d_m = \frac{m}{V}$$

$$\text{Mass density} = \frac{\text{mass}}{\text{volume}} \leftarrow \text{units} \frac{\text{grams}}{\text{meter}^3} \text{ or } \frac{g}{m^3}$$

$$\frac{\text{grams}}{\text{centimeters}^3} \text{ or } \frac{g}{cm^3}$$

$$d_w = \frac{w}{V}$$

$$\text{Weight density} = \frac{\text{weight}}{\text{volume}} \leftarrow \text{units} \left(\frac{\text{pounds}}{\text{foot}^3}\right) \text{ or } \left(\frac{Lb}{ft^3}\right)$$

These equations show that a lot of mass or weight contained in a small volume has a high density (see Table 2.2).

EXAMPLE 2.21

The density of water is 62.4 Lb/ft³. A 55-gal drum contains approximately 7.33 ft³ of volume (Fig. 2.20). How much does a 55-gal drum of water weigh?

$$d_w = 62.4 \, \frac{Lb}{ft^3}$$

$$V = 7.33 \, ft^3$$

$$w = ?$$

$$d_w = \frac{w}{V}$$

$$w = d_w \cdot V$$

$$w = \left(62.4 \, \frac{Lb}{ft^3}\right) (7.33 \, ft^3)$$

$$w = 458 \, Lb$$

Figure 2.20

Table 2.2

Table of Densities

Solid	g/cm³	Lbₘ/ft³
Aluminum	2.699	168.5
Aluminum, Al-Clad 17 ST alloy	2.96	185
Brass, yellow, cast	8.44	527
Brick (average)	1.9	120
Bronze, gun metal	8.78	548
Cement, dry	1.5	94
Cement, set	2.7–3.0	170–190
Concrete (average)	2.4	150
Copper, cast	8.30–8.95	518–559
Copper, hard-drawn	8.89	555
Douglas fir, dry	0.446	27.8
Glass, common	2.4–2.8	150–175
Gold, cast	19.3	1200
Ice (0°C)	0.917	57.2
Lead	11.3	705
Oak, white, dry	0.710	44.3
Paper	0.70–1.15	44–72
Pine, white, dry	0.373	23.3
Platinum	21.37	1334
Sand	1.5	94
Silver, cast	10.5	655
Solder, plumbing	9.4	590
Steel, 1 percent carbon	7.83	489
Steel, stainless	7.75	484
Tungsten	18.6–19.1	1160–1190
Tungsten carbide	14.0	874
Uranium	18.7	1170

*The temperature is 20°C [68°F] unless stated otherwise.

Note: 1 g/cm³ = 1 kg/L = 1000 kg/m³

EXAMPLE 2.22

Figure 2.21

Suppose you buy a mobile home, a double-wide trailer measuring 80.0 ft by 9.00 ft by 20.0 ft, and have it trucked to your lot (Fig. 2.21). If the density of air at the time of the move is 0.0810 Lb/ft³, what is the weight of the air being transported on the truck?

$$d_w = 0.0810 \frac{\text{Lb}}{\text{ft}^3}$$

$$V = l \times w \times h$$

$$= 80.0 \text{ ft} \times 9.00 \text{ ft} \times 20.0 \text{ ft}$$

$$= 14{,}400 \text{ ft}^3$$

$$w = ?$$

$$d_w = \frac{w}{V}$$

$$w = d_w \cdot V$$

$$= \left(0.0810 \, \frac{\text{Lb}}{\text{ft}^3} \right) (14{,}400 \, \text{ft}^3)$$

$$= 1{,}170 \, \text{Lb}$$

Who would ever think air could weigh so much!

 ## TIME

The standard measurement of time is the second. Time is measured by a clock. A *clock* is anything that repeats itself at a constant rate.

Atoms make for very accurate clocks. A cesium 133 atom, for example, emits microwave radiation at a very precise frequency when stimulated. A second is defined as the duration of 9,192,631,700 periods of oscillation of radiation coming from a cesium 133 atom. Listed below are a number of ways to measure time.

Time Measurement by Mechanical Devices

Pendulum   A pendulum could be simply a weight on a string swinging back and forth, repeating its motion at a regular rate. A repetitive motion is known as an *oscillation*. One back-and-forth motion is one oscillation. To time something with a pendulum, simply count the number of complete back-and-forth motions or oscillations during any action being timed and multiply this number by the time for one oscillation of the pendulum (Fig. 2.22).

Figure 2.22
A pendulum clock.

Figure 2.23

E X A M P L E   2.23

Make a pendulum from a weight tied onto a string 1 m long. If you pull the weight back a small amount, it will take 2 seconds to complete one oscillation for a pendulum of this length. Time your heart rate by counting the number of beats for 30 oscillations; this will give you your heart rate in beats per minute (Fig. 2.23).

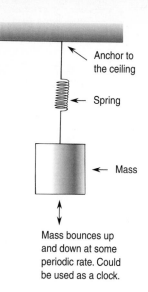

Figure 2.24
A spring clock.

Weight on a Spring   A weight on a spring, bouncing up and down, completes an oscillation at a regular rate. To time something, simply count the number of oscillations it takes for some action to be completed and multiply this by the time for one oscillation (Fig. 2.24).

Electronic Measurement Devices

RC Oscillator   The alternate charging and discharging of a capacitor through a resistor happens at a determined rate dependent on the time constant $\tau = R \times C$. This can be used to measure time. One of the electronic industry's standards that controls this charging and discharging is the 555-oscillator chip (Fig. 2.25). It has an accuracy of 1 percent.

EXAMPLE 2.24

An oscillator with the period of 1 ms may be used as a timer to cook a 1-minute, soft-boiled egg. Simply wait for 60,000 pulses, and the egg is done.

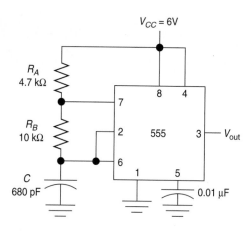

Figure 2.25
A 555 timer chip.

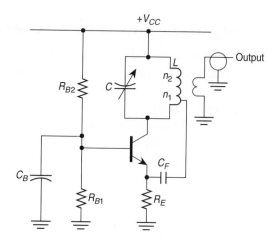

Figure 2.26
An LC oscillator.

LC Oscillator   This device has an accuracy 100 times better than the RC oscillator (Fig. 2.26). It has an accuracy of 0.01 percent.

Crystal Oscillator   This device is many times more accurate than the LC oscillator (Fig. 2.27). Over a normal temperature range, stabilities of a few parts per million are easily obtained.

Atomic Clock   Atomic clocks have the best accuracy of all the oscillators (Fig. 2.28). They currently have an accuracy of 1 part in 10^{12}.

Calibrating Clocks

As stated above, calibrating a measurement device means to adjust its output against a known standard so that it accurately displays what it is measuring. In the case of time, the best standard is an atomic clock. So, where do you get an atomic

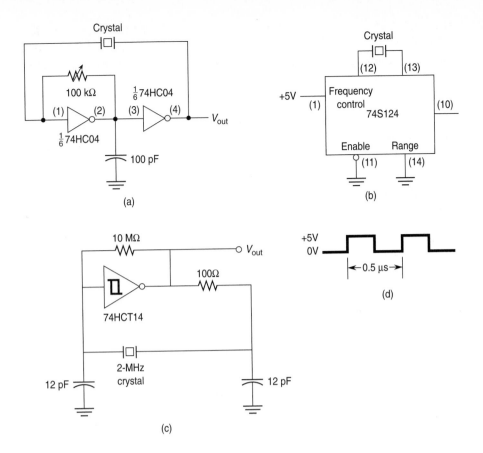

(a)

(b)

(c)

(d)

Figure 2.27
A crystal oscillator.

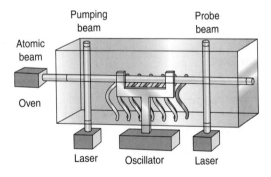

Figure 2.28
An atomic clock.

clock? The good news is that you don't need one. The National Institute of Standards and Technology (NIST) and radio station WWVB operate on short wave at 60 kHz and broadcast time and frequencies derived from their atomic clocks. This radio station can be used for calibration. For more information, look at the NIST Web site at www.boulder.nist.gov/timefreq.

CHAPTER SUMMARY

The bracket method is a systematic way to make unit conversions.

Step 1   Write down the conversions.

Step 2   Write down what is to be converted with a bracket next to it.

Step 3   Fill in the bracket with the conversion, placing the unit that is on the outside of the bracket in the bottom of the bracket and the unit it is equal to in the top part of the bracket.

When converting units that are squared or cubed, don't forget to raise the bracket to the same power.

Significant Digits

All nonzero numbers are significant. Zeros are significant if . . .

a) They have a bar placed over them.

b) They lie between significant digits.

c) They follow the decimal and a significant digit.

Adding and Subtracting Significant Digits
When quantities are added or subtracted, round the result to the decimal place of the number with the fewest decimal places in the addition or subtraction.

Multiplying and Dividing Significant Digits
When quantities are multiplied or divided, round the result so that it has no more significant digits than the number with the least significant digits in the multiplication or division.

Conversions
Length
$$1 \text{ in.} = 2.54 \text{ cm}$$
$$1 \text{ m} = 100 \text{ cm}$$
$$1 \text{ cm} = 10 \text{ mm}$$
$$1 \text{ mi} = 5{,}280 \text{ ft}$$
$$1 \text{ yd} = 3 \text{ ft}$$
$$1 \text{ ft} = 12 \text{ in}$$

Area
$$1 \text{ m}^2 = 10{,}000 \text{ cm}^2$$
$$1 \text{ ft}^2 = 144 \text{ in}^2$$

$$1 \text{ cm}^2 = 0.155 \text{ in}^2$$
$$1 \text{ ft}^2 = 929 \text{ cm}^2$$

Volume

SI
$$1 \text{ liter (L)} = 1{,}000 \text{ cm}^3$$
$$1 \text{ m}^3 = 10^3 \text{ liters}$$
$$1 \text{ cm}^3 = 1{,}000 \text{ mm}^3$$

English
$$1 \text{ gal} = 4 \text{ qt} = 8 \text{ pints} = 128 \text{ fl. oz.}$$
$$1 \text{ ft}^3 = 7.48 \text{ gal}$$

English to SI
$$1 \text{ ft}^3 = 28.3 \text{ liters}$$
$$25.3 \text{ ft}^3 = 1 \text{ m}^3$$
$$1 \text{ gal} = 3.785 \text{ liters}$$

$$d_m = \frac{\text{mass}}{\text{volume}} \quad \leftarrow \quad \text{mass density}$$

$$d_w = \frac{\text{weight}}{\text{volume}} \quad \leftarrow \quad \text{weight density}$$

Formulas for Area
Square: $A = a^2$

Rectangle: $A = a \times b$

Circle: $A = \pi r^2$

Right triangle: $A = 1/2(a \times b)$

Formulas for Volume
Cube: $V = a^3$

Rectangular volume: $V = l \times w \times h$

Cylinder: $V = \pi r^2 h$

Sphere: $V = \dfrac{4}{3} \pi r^3$

Problem-Solving Tips

1. Remember to raise the bracket in a conversion to the same power that is being converted.

2. Don't forget to apply the rules of addition/subtraction and multiplication/division of significant digits when calculating.

PROBLEMS

1. Convert 1.2 mi into feet.
2. Calculate the distance to the moon in feet. The moon is 250,000 mi away.
3. Convert 25 mi into feet.
4. Convert 3.5 yd into feet.
5. Convert 10 ft into inches.
6. Convert 12 in into centimeters.
7. Convert 5.6 m into centimeters.
8. Convert 25 cm into millimeters.
9. Convert 350 μm into millimeters.
10. Convert 0.8 yd into inches.
11. Convert 2.1 ft into centimeters.
12. Convert 25 in^2 into square centimeters.
13. Convert 59 in^3 into cubic meters.
14. Convert 327 in^3 into liters.
15. Calculate the area of a rectangle with a length of 3.0 m and a width of 5.0 m.
16. Calculate the area of a circle with radius of 1.5 cm.
17. Calculate the area of a right triangle with a base of 2.1 in and height of 4.8 in.
18. Calculate the surface area of a cylinder with a radius of 23 cm and a height of 4.9 cm.
19. Calculate the volume of a sphere with a radius of 5.6 mm.
20. A car has a 5.6-liter engine. How many cubic centimeters is this?
21. A pump fills a pool with 5,000 liters of water. How many cubic inches is this?
22. Calculate how many seconds there are in one year.
23. A man lives to be 72 years old. How many minutes is this?
24. A clock pulse from an oscillator is on for 50 microseconds and off for 25 microseconds. If 100 pulses have elapsed, how long has the clock been running?
25. A pendulum takes 1.2 seconds to complete one oscillation. If the pendulum completes 32 oscillations, how much time has elapsed?
26. A strain gauge is used to measure the weight of some object. The gauge emits a signal of 10 Lb/mV. If the gauge is measuring 25 mV, how much does the object weigh?
27. An object has a density of 84.5 g/cm^3 and a mass of 13.2 g. What is its volume?
28. A radar gun measures the speed of a moving car by determining the difference in frequency of the outgoing and incoming waves it emits. The signal out of the gun is equal to $\dfrac{10\frac{mi}{h}}{Hz}$. Suppose the gun measures a frequency difference of 9 Hz. How fast is the car moving in miles per hour?

Taking into account the rules for calculating with significant digits, calculate the following.

29. 3.21 in + 3.1 in + 4.675 in = ?
30. 2.76×10^3 m − 2.4×10^2 m = ?
31. (6.67 in) (2.1 in) (3.32 in) = ?
32. $\dfrac{0.012 \text{ m}}{0.00354 \text{ m}}$ = ?
33. What is the density of a ball with a radius of 2 in and weighing 1.7 Lb?

Linear Motion

L*inear motion* means movement in a straight line. The equations that describe this motion are used in the programs that run elevators, computer numerical control (CNC) milling machines, military ballistics, and robotic arms that put together anything from automobiles to electronic circuit boards. This chapter will describe how to measure and calculate this motion.

 RATES

Figure 3.1

A *rate* is a measure of how something changes over time. For example, the rate of water flowing out of a faucet is measured in gallons per minute (gal/min), and the rate of a machine gun firing is measured in rounds per second (rounds/second). Rates have units of some quantity over time (quantity/time). Speed is also a rate. *Speed* is a measure of how far something travels in a given amount of time. If a car travels 50 mi in 1 hour, then the average speed of the car is 50 mi/h.

 AVERAGE SPEED

A car driving around town may have changed its speed many times, slowing down for a red light and speeding up after the light turns green. This average speed $\bar{v}$ can be calculated by:

$$\bar{v} = \frac{d}{t}$$

$$\text{Average speed} = \left(\frac{\text{distance}}{\text{time}}\right)$$

 VELOCITY

Velocity is similar to speed, but it also includes a direction of travel.[1]

---

[1] Note that velocity and speed can sometimes have different values. For example, a boy running up and down a basketball court can have an average speed of 5 miles per hour, but because he is running in opposite directions every other lap, his velocity averages out to zero.

EXAMPLE 3.1

A car moves at an average speed of 50 mi/h, traveling northwest. The average velocity of the car is 50 mi/h, northwest.

 ## VECTOR

A quantity that has both magnitude and direction is called a *vector*. Velocity is an example of a vector. It has both magnitude or size and direction.

 ## SCALAR

A *scalar* is a quantity without a direction attached to it. Speed, mass, length, and time are scalar quantities.

 ## COORDINATE SYSTEMS

A typical coordinate system defines up as positive and down as negative, and to the right as positive and to the left as negative (Figure 3.2).

EXAMPLE 3.2

An elevator moving upward at a speed of 5 ft/s has a velocity of +5 ft/s (Fig. 3.3). If the same elevator is moving downward at a speed of 5 ft/s, its velocity is −5 ft/s. The sign indicates the direction. As you can see, velocity describes both speed and direction.

EXAMPLE 3.3

A robot travels 20 ft to the right in 4 seconds (Fig. 3.4). What is its average velocity?

$$d = 20 \text{ ft}$$

$$t = 4 \text{ s}$$

$$\bar{v} = ?$$

$$\bar{v} = \frac{d}{t}$$

$$\bar{v} = \left(\frac{20 \text{ ft}}{4 \text{ s}}\right)$$

$$\bar{v} = 5 \frac{\text{ft}}{\text{s}}, \text{ to the right}$$

$$\text{or } \bar{v} = +5 \frac{\text{ft}}{\text{s}}$$

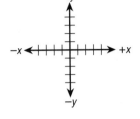

Figure 3.2
A typical coordinate system defines up as positive and down as negative, and to the right as positive and to the left as negative.

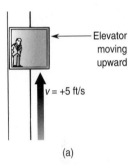

(a)

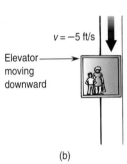

(b)

Figure 3.3

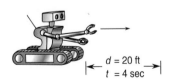

Figure 3.4

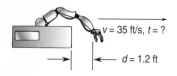

Figure 3.5

$v = 35$ ft/s, $t = ?$

← $d = 1.2$ ft

EXAMPLE 3.4

A robotic arm moving at an average speed of 35 ft/s travels 1.2 ft. How long did it take to travel this distance?

$$\bar{v} = 35\,\frac{\text{ft}}{\text{s}}$$

$$d = 1.2 \text{ ft}$$

$$t = ?$$

$$\bar{v} = \frac{d}{t}$$

$$t = \frac{d}{\bar{v}}$$

$$t = \frac{1.2 \text{ ft}}{35\dfrac{\text{ft}}{\text{s}}} \quad \text{Note that } \frac{1}{\left(\dfrac{1}{\text{s}}\right)} = \text{s}$$

$$t = 0.034 \text{ s}$$

EXAMPLE 3.5

The head of a CNC milling machine moves across the table at an average speed of 5 ft/s. How far did the head travel in 0.2 seconds?

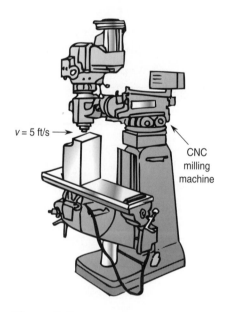

$v = 5$ ft/s →

CNC milling machine

Figure 3.6

$$\bar{v} = 5\,\frac{\text{ft}}{\text{s}}$$

$$t = 0.2 \text{ s}$$

$$d = ?$$

$$\bar{v} = \frac{d}{t}$$

$$d = \bar{v}t$$

$$d = \left(5\,\frac{\text{ft}}{\cancel{s}}\right)0.2\,\cancel{s}$$

$$d = 1.0\,\text{ft}$$

 ## INSTANTANEOUS VELOCITY

Instantaneous velocity is the velocity of an object at one point in time. For instance, the velocity of a car 5 seconds after the traffic light turns green.

We shall use the symbols for initial and final velocity:

$$v_i = \text{velocity}_{\text{initial}}$$
$$v_f = \text{velocity}_{\text{final}}$$

EXAMPLE 3.6

A race car starts from rest at a stoplight; after passing the finish line, the velocity of the car is +440 ft/s. What are the initial and final velocity of the car?

$$v_i = 0\,\frac{\text{ft}}{\text{s}}$$

$$v_f = +440\,\frac{\text{ft}}{\text{s}}$$

$v_i = 0$ $v_f = +440$ ft/s

Figure 3.7

EXAMPLE 3.7

A ball is tossed upward at 12 ft/s and reaches some maximum height before changing its direction and falling back down. What are the initial velocity and the final velocity of the ball when it reaches maximum height?

$$v_i = +12\,\frac{\text{ft}}{\text{s}}$$

$$v_f = 0\,\frac{\text{ft}}{\text{s}}$$ When the ball reached maximum height, it was not moving

$v_i = +12$ ft/s

Figure 3.8

PROBLEM-SOLVING TIP A projectile traveling upward reaches maximum height when it changes its direction of travel, and at this point its velocity is zero.

A nonelectronic and/or mechanical car speedometer measures the speed of the car using a steel cable connected to the transmission (Fig. 3.9a). The cable spins at a rate proportional to the car's speed. The cable enters a speed cup, which consists of a rotating magnet inside an aluminum cup attached to a spring and a needle. As the magnet spins, it induces an electric current in the cup, which produces its own magnetic field. The two magnetic fields interact, producing a torque in the cup, causing it to rotate the needle (Fig. 3.9b).

An electronic speedometer is shown in Fig. 3.10.

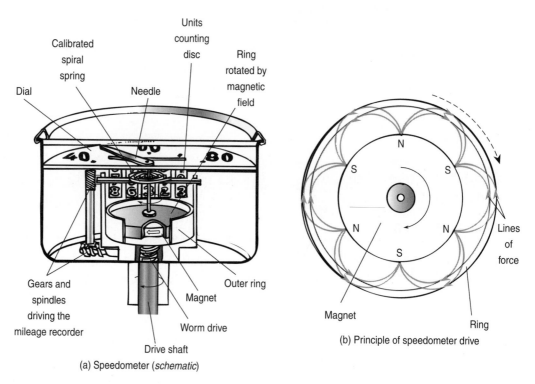

Figure 3.9
A mechanical car speedometer.

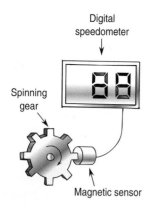

Figure 3.10
An electronic speedometer uses a magnetic sensor to determine the rate at which the transmission gears are spinning and converts this into a speed.

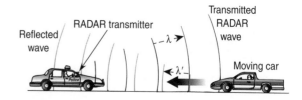

Figure 3.11
A police radar gun.

Radar Gun   A police radar gun consists of a microwave radio source emitting a beam of radio waves of a given frequency (Fig. 3.11). The waves are shot at moving vehicles. After reflecting off a car, the wave returns altered in frequency because of the car's motion. This is known as the *Doppler effect,* which will be studied in a later chapter. The reflected waves' frequency is compared with the frequency of the original wave sent out. The difference in the frequency of the two waves is a measure of how fast the car is moving.

ACCELERATION

Acceleration is how fast something changes its velocity.

Calculating Acceleration

Mathematically, acceleration is a change of velocity with time. Acceleration is the rate of a rate; therefore, it has units of $\dfrac{\text{length}}{\text{time}^2}$. The average acceleration is given by the formula:

$$\bar{a} = \frac{v_f - v_i}{t}$$

$$\text{average acceleration} = \frac{\text{velocity}_{\text{final}} - \text{velocity}_{\text{initial}}}{\text{time}}$$

TIP   To determine whether an object is slowing down or speeding up, determine:

- If a and v have the same sign $\rightarrow$ The object is speeding up
- If a and v have opposite signs $\rightarrow$ The object is slowing down (deceleration).

EXAMPLE 3.8

An object moving in a straight line, accelerating at $+25$ ft/s^2, means every second, the object increases its velocity $+25$ ft/s. If the object started from rest, after 1 second it would be moving at $+25$ ft/s; after 2 seconds, $+50$ ft/s; 3 seconds, $+75$ ft/s, and so on.

EXAMPLE 3.9

An elevator moves from zero to $+30$ ft/s in 5 seconds. What is its average acceleration?

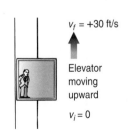

$v_f = +30$ ft/s

Elevator moving upward

$v_i = 0$

Figure 3.12

$$v_i = 0 \, \frac{\text{ft}}{\text{s}}$$

$$v_f = +30 \, \frac{\text{ft}}{\text{s}}$$

$$t = 5 \, \text{s}$$

$$\bar{a} = ?$$

$$\bar{a} = \frac{v_f - v_i}{t}$$

$$a = \frac{+30 \, \frac{\text{ft}}{\text{s}} - 0 \, \frac{\text{ft}}{\text{s}}}{5 \, \text{s}}$$

$$a = +6 \, \frac{\text{ft}}{\text{s}^2}$$

As you can see, acceleration has units of $\frac{\text{distance}}{\text{time}^2}$.

EXAMPLE 3.10

Figure 3.13

A machine moves across the floor, changing its velocity from $+11$ ft/s to $+3$ ft/s, and it takes 4 seconds to do this. What is its deceleration?

$$v_i = +11 \, \frac{\text{ft}}{\text{s}}$$

$$v_f = +3 \, \frac{\text{ft}}{\text{s}}$$

$$t = 4 \, \text{s}$$

$$\bar{a} = ?$$

$$\bar{a} = \frac{v_f - v_i}{t}$$

$$a = \frac{+3 \, \frac{\text{ft}}{\text{s}} - 11 \, \frac{\text{ft}}{\text{s}}}{4\text{s}}$$

$$= -2 \, \frac{\text{ft}}{\text{s}^2}$$

Acceleration Due to Gravity (Free-Fall Acceleration)

Drop an object. As the object falls, it picks up speed. This picking up of speed, or acceleration, is the acceleration due to gravity known as g.

$$g = -32.2 \text{ ft/s}^2$$
$$g = -9.8 \text{ m/s}^2 \text{ in the metric system}$$

Note: g is negative because it points downward.

In free-fall acceleration, g acts the same upon every object independent of its weight. For example, if a bowling ball and a marble were dropped at the same time, they both would hit the ground at the same time (Fig. 3.14). This is because gravity acts the same upon every object regardless of its weight; this is neglecting air resistance, of course.[2]

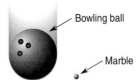

Figure 3.14
Both objects fall at the same rate and therefore hit the ground at the same time.

EXAMPLE 3.11

Take a penny and a quarter, and drop them at the same time. Do they hit the ground at the same time?

In all problems where an object is freely falling or being acted upon by gravity, acceleration $a = g$.

The free-fall acceleration can vary slightly over the earth's surface depending on what is underneath the surface. For example, if there is a large ore deposit underneath the surface, the acceleration due to gravity will be slightly enhanced because ore is denser than ordinary soil. Mining companies use this property to determine where to dig for ore and other metals (Fig. 3.16).

Quarter → ← Penny

Figure 3.15

Falling object

Ore deposit

Figure 3.16
Determining ore deposits.

---

[2] The reason everything falls at the same rate is because inertial mass and gravitational mass are equivalent. These concepts will be discussed in later chapters.

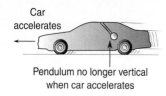

Figure 3.17
A pendulous accelerometer.

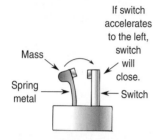

Figure 3.18
An inertia switch in an automobile shuts off the fuel supply in the event of a crash.

Accelerometers

A device that measures acceleration is called an *accelerometer.*

Pendulous Type   Pendulous-type accelerometers have a suspended mass, with some type of transducer measuring the position of the mass (Fig. 3.17). The mass will move only if accelerated while the transducer will pick up this motion.

Inertia Switch   Many cars now have an *inertia switch,* which measures acceleration (Fig. 3.18). If the acceleration is too large, such as in the event of a collision, the inertia switch shuts off the fuel pump and/or deploys the air bags.

Resistive Accelerometer   A *resistive accelerometer* is essentially a strain gauge with a mass mounted on it. During acceleration, the mass produces a stress on the resistive element. Stressing the resistor produces a change in its resistance (Fig. 3.19).

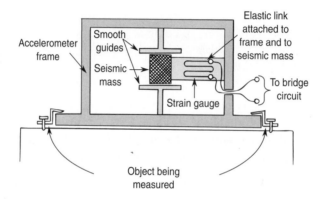

Figure 3.19
A resistive accelerometer. During acceleration, the mass produces a stress on the resistive element. Stressing the resistor produces a change in its resistance.

Accelerated Systems with a Constant Acceleration

If the acceleration is a constant (as in the case of gravity), then the average acceleration ($\bar{a}$) is the same as the instantaneous acceleration (a). Putting together the equations we've discussed, along with some other useful relations, we obtain:

For any type of motion:

$$\bar{v} = \frac{d}{t} \qquad \bar{a} = \frac{v_f - v_i}{t}$$

For accelerated motion with a constant acceleration:

$$\bar{v} = \frac{v_i + v_f}{2} \qquad d = \left(\frac{v_i + v_f}{2}\right) \cdot t \qquad v_f = v_i + a \cdot t$$

$$d = v_i \cdot t + \frac{1}{2} a \cdot t^2 \qquad 2 \cdot a \cdot d = v_f^2 - v_i^2$$

When a problem involves acceleration, time, and initial and final velocity, the following equation may be used:

$$v_f = v_i + a \cdot t$$

$$\text{Velocity}_{\text{final}} = \text{velocity}_{\text{initial}} + \text{acceleration} \times \text{time}$$

EXAMPLE 3.12

A load is dropped from a crane at rest and falls for 2.0 seconds. How fast in feet per second is the load traveling after 2 seconds? Remember, if anything is falling, it is because gravity is pulling on it, causing it to accelerate.

$$v_i = 0$$

$$t = 2.0 \; s$$

$$a = -32.2 \, \frac{ft}{s^2}$$

$$v_f = ?$$

$$v_f = v_i + a \cdot t$$

$$v_f = 0 + \left(-32.2 \, \frac{ft}{s^2}\right) \cdot 2.0 \; s$$

$$v_f = -64 \, \frac{ft}{s}$$

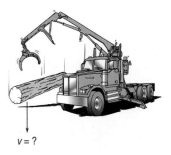

$v = ?$

Figure 3.20

When a problem involves acceleration, distance, time and velocity, the following equation may be used:

$$d = v_i \cdot t + \frac{1}{2} a \cdot t^2$$

$$\text{Distance} = \text{initial velocity} \times \text{time} + \frac{1}{2} \text{aceleration} \times \text{time}^2$$

EXAMPLE 3.13

A machine rolls across a platform, accelerating at a rate of a = +3.00 ft/s² with an initial velocity of +6.00 ft/s, traveling for a time of 1.50 seconds. How far does the machine travel?

$$v_i = +6.00 \, \frac{ft}{s}$$

$$a = +3.00 \, \frac{ft}{s^2}$$

$$t = 1.5 \; s$$

$$d = ?$$

$$d = v_i t + \frac{1}{2} a t^2$$

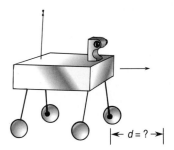

$d = ?$

Figure 3.21

$$d = \left(6.00\,\frac{\text{ft}}{\cancel{\text{s}}}\right) 1.50\,\cancel{\text{s}} + \frac{1}{2}\left(3.00\,\frac{\text{ft}}{\cancel{\text{s}^2}}\right)(1.50\,\cancel{\text{s}})^2$$

$$d = 12.4 \text{ ft} \quad \leftarrow \quad \text{Note: the seconds canceled out}$$

EXAMPLE 3.14

A man drops a rock off a bridge from rest, and it takes 4.0 seconds for the rock to hit the water below. In meters, how high is the bridge above the water? Remember, the rock is falling because of gravity (g = −9.8 m/s²).

$$v_i = 0\,\frac{\text{m}}{\text{s}} \quad \leftarrow \quad \text{The rock starts from rest}$$

$$a = -9.8\,\frac{\text{m}}{\text{s}^2}$$

$$t = 4.0 \text{ s}$$

$$d = ?$$

$$d = v_i t + \frac{1}{2}\bar{a}t^2$$

$$d = \left(0\,\frac{\text{m}}{\text{s}}\right)4.0\text{ s} + \frac{1}{2}\left(-9.8\,\frac{\text{m}}{\cancel{\text{s}^2}}\right)(4.0\,\cancel{\text{s}})^2$$

$$= -78 \text{ m} \leftarrow \text{Note: the seconds canceled out}$$

Figure 3.22

When problems involve motion with no reference to time, the following equations may be used:

$$\bar{v} = \frac{v_i + v_f}{2}$$

$$\text{Average velocity} = \frac{\text{velocity}_{\text{initial}} + \text{velocity}_{\text{final}}}{2}$$

or

$$2a \cdot d = v_f^2 - v_i^2$$

$$2 \times \text{acceleration} \times \text{distance} = (\text{velocity}_{\text{final}})^2 - (\text{velocity}_{\text{initial}})^2$$

EXAMPLE 3.15

What is the average velocity of an electron inside a CRT shot out of an electron gun at 6.0 × 10⁷ m/s and stopped by the screen?

$$v_f = 0$$

$$v_i = +6.0 \times 10^7\,\frac{\text{m}}{\text{s}}$$

$$\bar{v} = ?$$

$$\bar{v} = \frac{v_i + v_f}{2}$$

$$\bar{v} = \frac{6.0 \times 10^7 \frac{m}{s} + 0}{2}$$

$$\bar{v} = 3.0 \times 10^7 \frac{m}{s}$$

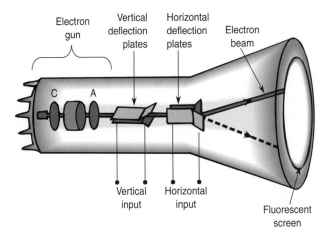

Figure 3.23

EXAMPLE 3.16

A weight is dropped from a resting position from a pile driver, at a height of 25 ft, onto a pile below. How fast does the weight hit the pile?

$$d = -25 \text{ ft}$$

$$v_i = 0$$

$$a = -32.2 \frac{\text{ft}}{\text{s}^2}$$

$$v_f = ?$$

$$2\bar{a}d = v_f^2 - v_i^2$$

$$2\bar{a}d + v_i^2 = v_f^2 - \cancel{v_i^2} + \cancel{v_i^2} \quad \leftarrow \quad \text{Solving for } v_f$$

$$v_f^2 = 2\bar{a}d + v_i^2$$

$$v_f = \sqrt{2\bar{a}d + v_i^2}$$

$$v_f = \sqrt{2\left(-32.2 \frac{\text{ft}}{\text{s}^2}\right)(-25\text{ft}) + (0)^2}$$

$$= 40 \frac{\text{ft}}{\text{s}}$$

Figure 3.24

EXAMPLE 3.17

A bullet is shot straight up into the air. The bullet reaches a maximum height of 3000 ft. What was its initial velocity?

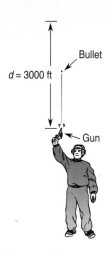

Figure 3.25

$$d = 3\bar{0}00 \text{ ft}$$

$$a = -32.2 \frac{\text{ft}}{\text{s}^2}$$

$$v_f = 0 \quad \leftarrow \quad \text{At maximum height the bullet is still}$$

$$v_i = ?$$

$$2\bar{a}d = v_f^2 - v_i^2$$

$$2\bar{a}d + v_i^2 = v_f^2 - \cancel{v_i^2} + \cancel{v_i^2} \quad \leftarrow \quad \text{Solving for } v_i$$

$$2\bar{a}d + v_i^2 = v_f^2$$

$$2\bar{a}d - 2\bar{a}d + v_i^2 = v_f^2 - 2\bar{a}d$$

$$v_i^2 = v_f^2 - 2\bar{a}d$$

$$v_i = \sqrt{v_f^2 - 2\bar{a}d}$$

$$v_i = \sqrt{0^2 - 2\left(-32.2 \frac{\text{ft}}{\text{s}^2}\right)(3\bar{0}00 \text{ ft})}$$

$$= \sqrt{193200 \frac{\text{ft}^2}{\text{s}^2}}$$

$$= +440 \frac{\text{ft}}{\text{s}}$$

How do you determine which equation to use when solving a problem?

Problem-Solving Procedure

1. Make a formula sheet and have it in front of you.
2. As you read through the problem, write down the values of the quantities given.
3. Make a sketch of the problem, labeling everything to help you visualize what is being asked.
4. Work the problem in a vertical column, organizing the information and writing down everything that is relevant.
5. List all the relevant equations. If you have a problem figuring out which formulas apply, you can use the process of elimination to narrow down the possibilities. For example, if there is no mention of time in the problem, cross out all formulas dealing with time.
6. Work out any algebra first before plugging in numbers.
7. Make sure the quantities have the correct units when you are plugging them into formulas. If working in the mks or fps systems, stay in the systems when plugging in.
8. Check to see if the answer makes sense both numerically and unit-wise.

EXAMPLE 3.18

An air bag is deployed from rest during an accident at +540 m/s. What is the average velocity of the air bag?
    Follow the steps above to solve the problem.

1. Have a formula sheet in front of you.

Air bag deployed

$v_f = +540$ m/s
$v_i = 0$ $\bar{v} = ?$

Figure 3.26

For any type of motion: $\bar{v} = \dfrac{d}{t}$ $\qquad \bar{a} = \dfrac{v_f - v_i}{t}$

For accelerated motion with a constant acceleration:

$$\bar{v} = \frac{v_i + v_f}{2} \qquad d = \left(\frac{v_i + v_f}{2}\right) \cdot t \qquad v_f = v_i + a \cdot t$$

$$d = v_i \cdot t + \frac{1}{2}a \cdot t^2 \qquad 2 \cdot a \cdot d = v_f^2 - v_i^2$$

2. As we read the problem, we first read "from rest," which means $v_i = 0$. Then we read "at +540 m/s," which means $v_f = +540$ m/s. Finally, we read "What is the average velocity?" which means $\bar{v} = ?$

3. Draw a sketch.

4. Work the problem in a vertical column, writing everything down.

$$v_f = +540 \text{ m/s}$$

$$v_i = 0 \text{ m/s}$$
$$\bar{v} = ?$$

5. List all the relevant equations. If you have a problem determining which formulas apply, you can use the process of elimination to narrow down the possibilities. For example, in this problem, there is no mention of time, so cross out all formulas dealing with time.

$$\bar{v} = \frac{d}{t}$$

$$\bar{a} = \frac{v_f - v_i}{t}$$

$$\bar{v} = \frac{v_i + v_f}{2}$$

$$d = \left(\frac{v_i + v_f}{2}\right) \cdot t$$

$$v_f = v_i + a \cdot t$$

$$d = v_i \cdot t + \frac{1}{2}a \cdot t^2$$

$$2 \cdot a \cdot d = v_f^2 - v_i^2$$

From these remaining equations, only one fits our needs: $\bar{v} = \dfrac{v_i + v_f}{2}$

6. Work out the algebra before plugging in the numbers.

$$v_f = +540 \text{ m/s}$$

$$v_i = 0 \text{ m/s}$$
$$\bar{v} = ?$$

$$\bar{v} = \frac{v_i + v_f}{2}$$

Work out the algebra. No algebra needed here.

7. Plug in the numbers, checking first to see that all units match. Everything is in mks.

$$\bar{v} = \frac{0 + 540 \text{ m/s}}{2} = +270 \text{ m/s}$$

8. $\bar{v}$ should have units of meters per second, and it does.

CHAPTER SUMMARY

Anything that moves in industry—from robotic arms to elevators, conveyor belts, or air bags in a car—can be described by the equations in this chapter. The problem set at the end of the chapter contains relevant issues you may encounter when working in industry.

For any type of motion: $\bar{v} = \dfrac{d}{t}$ $\bar{a} = \dfrac{v_f - v_i}{t}$

For accelerated motion with a constant acceleration:

$$\bar{v} = \frac{v_i + v_f}{2} \quad d = \left(\frac{v_i + v_f}{2}\right) \cdot t \quad v_f = v_i + a \cdot t$$

$$d = v_i \cdot t + \frac{1}{2} a \cdot t^2 \qquad 2 \cdot a \cdot d = v_f^2 - v_i^2$$

Problem-Solving Tips

- As you read through the problem, write down the values of the quantities given.

- Have all your formulas in front of you and choose only the relevant ones. You can use the process of elimination to choose the formulas that are useful.

For example, if a problem does not mention time, then time is not relevant to the solution; therefore, choose equations without t. From the remaining equations, determine the proper one from the context of the problem.

- If something is at rest, its velocity is zero.
- When a projectile reaches maximum height, it changes direction, and $v = 0$.
- Pay close attention to the units, remembering to stay in either the metric or the English system of units.

PROBLEMS

1. A robotic arm moves a distance of 3 m in 2 seconds. What is the average speed of the arm?

2. A radio transmission is broadcasting from an antenna. The radio wave travels for 3 seconds. How far has the wave traveled? Note: Radio waves travel at the speed of light, at a constant of 3×10^8 m/s.

3. An elevator travels upward three floors at an average speed of 0.5 floors/s. How long will it take to travel 3 floors?

4. An ultrasonic tape measure is used to measure the distance across a room. It takes 0.1 seconds for the sound wave to travel to the opposite wall and back. How far away is the wall? Note: The speed of sound at 72°F is 330 m/s.

5. A car starts from rest and accelerates for 10 seconds, reaching a speed of 100 ft/s. What is

the acceleration of the car? How does this compare to the acceleration due to gravity?

6. An accelerometer is placed on a plane and emits a signal of 10 mV/(m/s²). At takeoff, the accelerometer measures 25 mV. What is the acceleration of the plane? Hint: Let the units guide you. Add, subtract, multiply, or divide to end up with meters per second squared in the answer.

7. Laser ranging is a method used to determine the distance to an object. A laser pulse is sent out, and a clock records how long it takes the pulse to return after reflecting off of something. This information, along with the speed of light, is used to determine the distance to the object. Suppose a pulse is sent out and takes 2 microseconds to return. Given that the speed of light is 3×10^8 m/s, how far away is the object?

8. An arrow is shot straight up into the air and reaches a height of 300 m. What was its initial velocity?

9. A conveyer belt transports parts on an assembly line. The conveyer belt starts from rest and accelerates at 1.2 ft/s^2 for a time of 4 seconds. How far did a part travel on the belt?

10. An electronic signal travels at 90 percent of the speed of light through a wire. If the signal travels for 1 nanosecond, how far does it travel? Note that the speed of light is 3×10^8 m/s.

11. You are in a car traveling at 55 mi/h, and you step on the gas, accelerating at a rate of 600 mi/h^2. How fast are you traveling after 1 minute?

12. Suppose the speedometer on your car broke and you wanted to determine the speed of your car on the highway. How would you use the mile markers on the roadside to determine your speed?

13. What is the difference between acceleration and velocity?

14. Which falls faster when dropped, a penny or a quarter? How fast do these objects pick up speed as they fall?

15. How might you build an accelerometer using chewing gum and some string?

Force and Momentum

This chapter describes force and momentum and how they are related to each other. It will look at different forms of measurement and how to calculate with them. The concepts learned in this chapter, along with the equations of motion learned in the previous chapter, will provide a more complete understanding of the mechanics of motion.

FORCE

A *force* is a push or pull in a certain direction (Figs. 4.1 and 4.2).

Figure 4.1
Your weight is an example of a force. Gravity is pulling you down toward the ground.

Figure 4.2
An elevator moves upward because a force below (a hydraulic cylinder) pushes it upward.

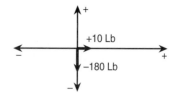

Figure 4.3
Instead of saying up or down, or to the left or to the right, we can use a coordinate system to designate direction.

Force is a vector because it has both a magnitude and direction. For example, force = 180 Lb downward or 10 Lb to the right. Instead of saying up or down, or to the left or to the right, we can use a coordinate system to designate direction (Fig. 4.3).

Measuring Force

A simple force gauge can be made from a spring (Fig. 4.4). In order to stretch or compress a spring, a force is needed. The amount the spring changes length is determined by the force.

Electronic Scales

Electronic scales typically use a strain gauge to measure weight or force. Most strain gauges are resistive. A resistive strain gauge consists of a very thin resistor mounted on a material that will flex as force is applied. A force applied to the resistor will deform it slightly, resulting in a change in its resistance (Fig. 4.5).

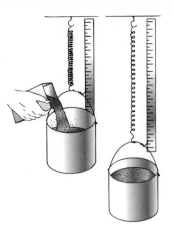

Figure 4.4
A spring force gauge.

Calibrating a Strain Gauge

Calibrating a sensor means to compare its output to a known input. To calibrate a strain gauge, apply weights of known value and record the output from the gauge. For example, suppose for the given set of weights, the output from the gauge was as follows:

Weight	Output Resistance
1.0 lb	10.8 kΩ
2.0 lb	9.2 kΩ
3.0 lb	7.6 kΩ
4.0 lb	6.0 kΩ

Plotting these data and drawing the best line through the points (fit to the data) gives us the following graph in Figure 4.6.

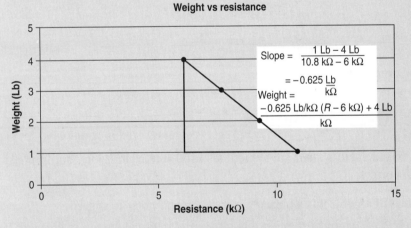

Weight vs resistance

$$\text{Slope} = \frac{1\ \text{Lb} - 4\ \text{Lb}}{10.8\ \text{k}\Omega - 6\ \text{k}\Omega}$$

$$= -0.625\ \frac{\text{Lb}}{\text{k}\Omega}$$

$$\text{Weight} = \frac{-0.625\ \text{Lb/k}\Omega\ (R - 6\ \text{k}\Omega) + 4\ \text{Lb}}{\text{k}\Omega}$$

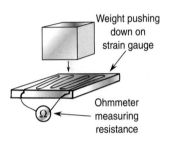

Figure 4.5
A resistive strain gauge converts an applied force, which deforms the gauge, into a change in resistance. The resistance is then converted into the corresponding weight.

Figure 4.6
A graph of weight vs. resistance along with a fit to the data.

We can use this fit to the data to generate an equation to model this sensor. Since the data fall on a straight line, we can therefore use an equation of a straight line to model it. Every straight line can be represented by the equation:

$$y = m \cdot x + b$$

where m is the slope

$$m = \frac{\text{rise}}{\text{run}}$$

$$b = \text{the } y\text{-intercept}$$

If the x-coordinate does not begin at zero but at x_0, we can write the equation as:

$$y = m(x - x_0) + b$$

In the case of the graph above:

$$m = \frac{1 \text{ Lb} - 4 \text{ Lb}}{10.8 \text{ k}\Omega - 6.0 \text{ k}\Omega} = -0.625 \frac{\text{Lb}}{\text{k}\Omega}$$

$$x_0 = 6.0 \text{ k}\Omega$$

$$b = 4.0 \text{ Lb}$$

Therefore, the equation of the line governing this sensor is:

$$\text{Weight} = -0.625 \frac{\text{Lb}}{\text{k}\Omega} \cdot (R - 6.0 \text{ k}\Omega) + 4.0 \text{ Lb}$$

We can then use this equation to determine the weight of any unknown weight in this range from the sensor's resistance.

EXAMPLE 4.1

A strain gauge is used to measure a force. Its resistance changes by 100 ohms per pound. What is the change in resistance when measuring 0.09 Lb? This problem is effectively asking for a conversion from pounds to ohms.[1]

$$100 \text{ ohms} = 1 \text{ Lb}$$

$$0.09 \text{ L̶b̶} \left(\frac{100 \text{ ohms}}{1 \text{ L̶b̶}} \right) = 9 \text{ ohms}$$

MASS AND WEIGHT

The words *mass* and *weight* are sometimes used interchangeably, but in physics this is not correct.

- The *mass* of an object is a measure of how much matter it contains.
- The *weight* of an object is a measure of the gravitational attraction the earth has for that object.

Mass and weight are related to each other by a constant, g (free-fall acceleration). If an object is in outer space, it has hardly any weight, but it still has mass. The mass of an object is the same everywhere because the amount of matter contained

---

[1] We know, of course, that it does not make sense in terms of dimensional analysis to convert pounds into ohms, but we can think of them as effectively equal in the context of this problem.

in it stays the same, but weight is gravity-dependent, that is, an astronaut weighs less on the moon because gravity is smaller (Fig. 4.7).

Just as we have units of length and time, we have units of mass and force. In the English system of units, force is measured in pounds and mass in a unit called slugs. In the SI system, mass is measured in kilograms, and force is measured in a unit called newtons.

	English System	SI System
Force	Pounds (Lb)	Newtons (N)
Mass	Slugs (slugs)	Kilograms (kg)

Figure 4.7
An astronaut has the same mass in outer space as on earth but has very little weight because gravity is so small there.

Conversions

1 kg = 1,000 g	1 g is about the mass of a paper clip
1 Lb = 4.45 N	1 newton is almost 1/4 Lb
1 slug = 14.6 kg	1 slug has the equivalent weight of 32.2 lb

To convert between mass and weight use:

$$\text{Weight} = \text{mass} \times \text{gravity}$$

Because weight is a force, we shall write:

$$F_w = m \cdot g \quad \leftarrow \quad g = 9.8 \, \frac{m}{s^2} \text{ metric or } g = 32.2 \, \frac{ft}{s^2} \text{ English}$$

Force due to weight = mass × acceleration due to gravity

EXAMPLE 4.2

An object has a mass of 10,000 g. What is its weight?

$$m = 10,000 \text{ g}$$
$$F_w = ?$$

We are working in the mks system; therefore, the mass must be converted into kilograms.

$$10,000 \text{ g} = 10 \text{ kg}$$

$$F_w = mg$$

$$F_w = (10 \text{ kg})\left(9.8 \, \frac{m}{s^2}\right) \quad \leftarrow \quad \text{Must use the metric acceleration due to gravity}$$

$$F_w = 98 \text{ N}$$

EXAMPLE 4.3

An object has a mass of 10 slugs. What is its weight?

$m = 1\bar{0}$ slugs

$F_w = ?$

$F_w = mg$

$F_w = (1\bar{0} \text{ slugs})\left(32.2\dfrac{\text{ft}}{\text{s}^2}\right)$ ← Must use the English acceleration due to gravity

$F_w = 320$ Lb

NEWTON'S LAWS

Newton's first law says that if a body changes its motion from standing still to moving, or changes from one speed to another, or changes its direction of travel, it is because there are forces acting upon it that do not balance each other out. For example, in a game of tug-of-war, if each side pulls with equal force, then the flag does not move; however, if one side pulls with more force, there will be a change in the motion of the team members (Fig. 4.8).

Figure 4.8
If the forces do not balance each other, there will be an acceleration of team members.

   Newton's second law says that in order to change an object's motion, a net force has to be applied to it. How fast you change its motion depends on how big the net force is. If the object is very massive and at rest, you have to apply a lot of force to move it. If an object is moving, to slow it down, speed it up, or change its direction, a force is needed. This statement can be summarized by the equation

$$a = \frac{\sum\limits_{i} F_i}{m}$$

$$\text{Acceleration} = \frac{\text{sum of the forces}}{\text{mass}}$$

This equation says that if something is being accelerated it is because there is a net force acting on it, and how much it accelerates depends on its mass and the applied force.

We can rewrite the above force equation as

$$\sum_i F_i = m \times a \quad \leftarrow \quad \text{Newton's second law}$$

Sum of forces = mass × acceleration

EXAMPLE 4.4

If a car full of people has a mass of 1,600 kg and accelerates at $+2.0\,\frac{m}{s^2}$ to the right, what is the net force acting on the car (Fig. 4.9)?

Figure 4.9
A force causes the car to accelerate.

$$m = 1,600 \text{ kg}$$

$$a = +2.0\,\frac{m}{s^2}$$

$$F = m \cdot a$$

$$F = (1,600 \text{ kg})\left(2.0\,\frac{m}{s^2}\right)$$

$$F = 3,200 \text{ N to the right}$$

or

$$F = +3,200 \text{ N}$$

EXAMPLE 4.5

A man pushes a crate on wheels that has a combined mass of 5.0 slugs with a force of 25 Lb. What is the acceleration of the crate?

$$m = 5.0 \text{ slugs}$$
$$F = 25 \text{ Lb}$$
$$a = ?$$
$$a = \frac{F}{m}$$
$$a = \frac{25 \text{ Lb}}{5.0 \text{ slugs}}$$
$$a = 5.0\,\frac{ft}{s}$$

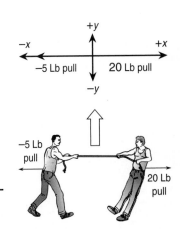

Figure 4.10
To represent force up, down, to the right or left, we will illustrate them this way.

 MULTIPLE FORCES

When dealing with problems involving more than one force, a force diagram is used to help clarify the problem (Fig. 4.10).

EXAMPLE 4.6

In a game of tug-of-war, two teams pull on a rope in the opposite directions, trying to pull the opposing team across a line. Suppose the members of the team on the left pull with forces 50 lb, 60 lb, and 45 lb, and members on the right team pull with forces of 35 lb, 70 lb, and 55 lb. What is the net force on the rope, and which way will it move?

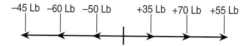

Figure 4.11

$$F_{\text{left}} = -50 \text{ Lb} - 60 \text{ Lb} - 45 \text{ Lb}$$
$$= -155 \text{ Lb}$$
$$F_{\text{right}} = 35 \text{ Lb} + 70 \text{ Lb} + 55 \text{ Lb}$$
$$= 160 \text{ Lb}$$

$$F_{\text{total}} = \sum_i F_i = F_{\text{right}} + F_{\text{left}}$$
$$= 160 \text{ Lb} - 155 \text{ Lb}$$
$$= +5 \text{ Lb} \leftarrow 5 \text{ Lb to the right. The rope will move to the right.}$$

EXAMPLE 4.7

An elevator weighing 2,100 Lb is being pulled upward with a force of 3,500 Lb. a) What is the total force acting on the elevator? b)What is its acceleration?

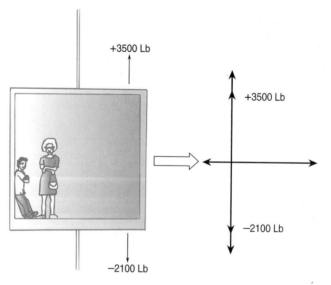

Figure 4.12
In our coordinate system, up is positive, and down is negative.

a)

$$F_{\text{up}} = +3,500 \text{ Lb}$$
$$F_{\text{down}} = -2,100 \text{ Lb}$$

$F_{\text{total}} = ?$

$F_{\text{total}} = F_{\text{up}} + F_{\text{down}}$

$F_{\text{total}} = 3{,}500\,\text{Lb} - 2100\,\text{Lb}$

$F_{\text{total}} = +1{,}400\,\text{Lb}$ ← A positive force means it is being accelerated upward.

b)

$$F_w = -2100\,\text{Lb}$$

$$F_{total} = +1400\,\text{Lb}$$

$$a = ?$$

$$m = \frac{F_w}{g}$$

$$m = \frac{-2100\,\text{Lb}}{-32.2\,\dfrac{\text{ft}}{\text{s}^2}}$$

$$m = 65\text{ slugs}$$

$$a = \frac{\sum_i F_i}{m}$$

$$a = \frac{+1{,}400\,\text{Lb}}{65\text{ slugs}}$$

$$a = +22\,\frac{\text{ft}}{\text{s}^2}$$

STATIC EQUILIBRIUM

Static equilibrium means all forces are counterbalanced by other forces, and nothing is accelerating. In other words, the static equilibrium means that:

$$\sum_i F_i = 0$$

The sum of all the forces pointing up equals the sum of all the forces pointing downward.

The sum of all the forces pointing to the right equals the sum of all the forces pointing to the left.

Man's weight downward

Chair pushes upward

Figure 4.13
The chair and the man are in static equilibrium.

EXAMPLE 4.8

A chair used to support a man weighing 165 lb must push upward with 165 Lb; otherwise it will collapse under his weight (Fig. 4.13).

EXAMPLE 4.9

A bridge is designed to support dynamic loads, that is, loads that change in time (Fig. 4.14). If a truck weighing 1 ton crosses the bridge, the bridge must exert a force of 1 ton upward, plus the force necessary to support itself, otherwise the truck will fall through the bridge.

Figure 4.14
A bridge is designed to support dynamic load.

FORCES ACTING IN TWO OR THREE DIMENSIONS

So far all the forces we have encountered have been acting in one dimension, (i.e., all in a straight line, either up and down or left and right). When there is a combination of forces acting in two or three dimensions, life gets more complicated, and trigonometry will be needed to solve such problems.[2] To solve these problems, the forces need to be broken up into x, y, and z components along the coordinate axes and then summed to determine the total force in each direction. This is demonstrated below.

EXAMPLE 4.10

Suppose your car gets stuck in the mud, but that doesn't bother you because you know physics! You tie a cable to the bumper and to a tree and stretch it tight. You apply a force of 55 Lb to the middle of the cable, creating a 5-degree angle with the horizontal. What tension is generated in the cable?

The problem is simplified using a force diagram (see Fig 4.15). To solve the problem, look at every force on the diagram and break it up into components along the x-axis and y-axis. If nothing is moving, we must have static equilibrium, that is:

Sum of the forces up = sum of the forces down

Sum of the forces to the left = sum of the forces to the right

[2] For a review of trigonometry see the appendix.

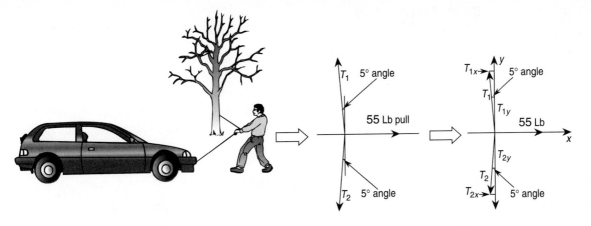

Figure 4.15
The problem can be simplified with a force diagram.

The applied force is 55 Lb.

$$T_3 = 55 \text{ Lb}$$
$$\theta = 5^0$$
$$T_1 = ?$$
$$T_2 = ?$$

By symmetry $T_1 = T_2$ and

$$T_{1y} = T_{2y} \quad \text{and} \quad T_{1x} = T_{2x}$$

Therefore, the sum of the forces to the left is:

$$T_{1x} + T_{2x} = T_{1x} + T_{1x} = 2T_{1x}$$

The sum of the forces to the left equals the forces to the right.

$$2T_{1x} = 55.0 \text{ Lb}$$

$$T_{1x} = \frac{55.0 \text{ Lb}}{2} = 27.5 \text{ Lb} \leftarrow \text{Solving for } T_{1x}$$

$$\sin \theta = \frac{\text{opposite}}{\text{hypotenuse}} \quad \text{Therefore}$$

$$\sin \theta = \frac{T_{1x}}{T_1} \quad \text{or}$$

$$T_1 = \frac{T_{1x}}{\sin \theta}$$

Plugging in gives $\rightarrow$

$$T_1 = \frac{27.5 \text{ Lb}}{\sin (5^0)} = 316 \text{ Lb}$$

Therefore, since $T_1 = T_2$

$$T_2 = 316 \text{ Lb}$$

Wow! You generated 316 Lb with 55 Lb. How is this possible? Why should the tension be higher than 55 Lb in two cables? The reason is because the two cables are both pulling not only to the right but also against each other, generating this higher tension.

The force of friction between two surfaces arises because the surfaces tend to dig into one another (Fig. 4.16).

The small block will dig in deeper than the big block because the weight is focused on a smaller area. The big block, however, has more peaks and valleys to hold on with. These two results balance each other out, making the force of friction independent of the surface area.

Key Points about Frictional Force

1. The force of friction between two surfaces is independent of the size of the two surfaces.
2. The force of friction between two surfaces is dependent on the quality of the surfaces.
3. The force of friction between two surfaces is dependent on the normal or perpendicular force to the surface. (If the surfaces are horizontal, the normal force is just the weight.)

These three statements can be summarized as:

$$F_f = \mu \cdot F_N$$

Frictional force = coefficient of friction × normal force

The coefficient of friction is a number without a unit. It is a measure of the slipperiness of two surfaces in contact. The normal force is the force perpendicular to the surface (Fig. 4.17).

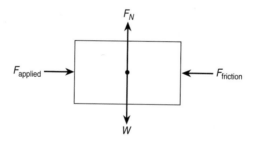

Figure 4.16
A magnified view of two surfaces in contact.

Figure 4.17
Frictional force acts in an opposite direction to the applied force.

Figure 4.18
In order to push the book, the frictional force between the book and the table needs to be overcome.

EXAMPLE 4.11

How much force is needed to begin to slide a book across a table when the book weighs 9.5 Lb and the coefficient of friction between the book and the table is $\mu = 0.2$? See Fig. 4.18.

$$F_N = 9.5 \text{ Lb}$$

$$\mu = 0.20$$

$$F_f = \mu F_N$$

$$F_f = (0.20)(9.5\ \text{Lb})$$

$$F_f = 1.9\ \text{Lb}$$

It takes more force to initially budge something than it does to sustain its motion (Fig. 4.19). This is because once the object is moving, the surfaces don't dig into each other as much. Therefore, the coefficient of starting or static friction (μ_s) is greater than the coefficient of sliding or kinetic friction (μ_k). See Table 4.1 for various coefficients.

$$\mu_s > \mu_k$$

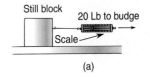

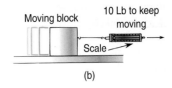

Figure 4.19
It takes more force to initially budge something than to keep it moving.

Table 4.1
Coefficients of Friction (μ)

Material	μ_S Starting Friction	μ_K Sliding Friction
Steel on steel	0.58	0.20
Steel on steel (lubricated)	0.13	0.13
Glass on glass	0.95	0.40
Hardwood on hardwood	0.40	0.25
Steel on concrete		0.30
Aluminum on aluminum	1.9	
Rubber on dry concrete	2.0	1.0
Rubber on wet concrete	1.5	0.97
Aluminum on wet snow	0.4	0.02
Steel on Teflon	0.04	0.04

Values listed are for ordinary pressures and temperatures, but may still vary somewhat from sample to sample.

EXAMPLE 4.12

How much force is required to budge a 100-Lb wooden crate on a wooden floor compared with the force needed to sustain its movement?

$$F_N = 100\ \text{lb}$$

$$\mu_s = 0.40$$

$$\mu_k = 0.25$$

Static case:

$$F_f = \mu_s F_N$$

$$F_f = (0.40)(100\ \text{Lb})$$

$$F_f = 40\ \text{Lb} \leftarrow \text{Force needed to budge the crate}$$

Kinetic case:

$$F_f = \mu_k F_N$$

$$F_f = (0.25)(100\ \text{Lb})$$

$$F_f = 25\ \text{Lb} \leftarrow \text{Force needed to keep the crate moving}$$

Figure 4.20
Improperly inflated tires add to rolling friction.

Rolling Friction

A rolling tire has friction because it is deformed as it rolls. A lot of energy goes into deforming the tire. A tire that is almost flat has a much harder time rolling than a properly inflated tire (Fig. 4.20). A steel wheel riding on steel, such as a train wheel, has almost no rolling friction because the wheel hardly deforms.

Reducing Friction

- Polishing surfaces or using lubricants can reduce friction. Too much polishing, however, can actually increase friction, such as in the case of two pieces of glass, which have difficulty sliding over one another.

- Using lubricants acts to fill in the voids in the surfaces that are in contact (Fig. 4.21). Two surfaces with a lubricant between them will tend to float over one another.

Figure 4.21
Lubricants act to fill in the voids in the surfaces in contact

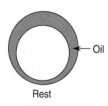

Rest

Starting

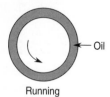

Running

Figure 4.22
Lubrication between surfaces.

- If the pressure between the two surfaces is great, a lubricant may have to be pumped between them (Fig. 4.22). Examples of this are in an engine on the piston walls, valve stems, and on the crankshaft and cam.

- Ball and roller bearing can further reduce friction by turning sliding friction into rolling friction (Fig. 4.23). The use of these bearings is sometimes necessary when there is no flow of lubricant possible, such as in a wheel.

Figure 4.23
Bearings convert kinetic friction into rolling friction.

Friction is normally thought of as something to be reduced, as in an engine. Friction is, however, a necessary thing. Think of how dangerous it can be driving on a wet road or how slippery a bathtub is when wet. In these cases, we want to increase friction for safety reasons (Fig. 4.24). In fact, if it were not for fingerprints increasing friction, it would be difficult to pick up anything (Fig. 4.25).

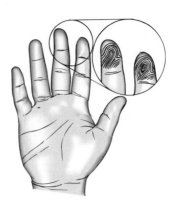

Figure 4.24
In the winter, some people place weight in the back of their pickup trucks to enhance traction on the slippery roads. The increase in weight will increase the friction between the tires and the road.

Figure 4.25
Fingerprints increase friction to aid in picking up objects.

MOMENTUM

If something is massive, it is difficult to move. Likewise, if something is massive and moving, it is difficult to stop. It is much more difficult to stop a semitrailer moving at 2 mph than a baby carriage moving at that speed. The semi has more momentum (Fig. 4.26).

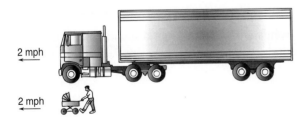

2 mph

2 mph

Figure 4.26
A semitrailer and a baby carriage are moving at the same speed. The semi is much more difficult to stop because it has a much greater momentum.

Mathematically, momentum is given by:

$$p = m \cdot v$$
$$\text{Momentum} = \text{mass} \times \text{velocity}$$

Something that is massive and moving fast has a lot of momentum. Momentum is a vector because it has both magnitude and direction. Since momentum is $m \times v$, it has units of:

$$\text{Momentum units: } \text{kg} \frac{\text{m}}{\text{s}} \text{ or slug} \frac{\text{ft}}{\text{s}}$$

EXAMPLE 4.13

$m = 890$ kg

←

+1.5 m/s

Figure 4.27

A car with a mass of 890 kg is moving at $+1.5$ m/s. What is its momentum?

$$m = 890 \text{ kg}$$

$$v = +1.5 \text{ m/s}$$

$$p = mv$$

$$p = (890 \text{ kg}) \left(1.5 \frac{\text{m}}{\text{s}} \right)$$

$$p = +1,300 \text{ kg} \frac{\text{m}}{\text{s}}$$

Momentum and force are related to each other by the equation:

$$F_{\text{total}} = \frac{\Delta p}{t} = \frac{p_f - p_i}{t}$$

$$\text{Total force} = \frac{\text{change in momentum}}{\text{time}}$$

This equation shows that if the momentum of something is changing in time, it is because there is a net nonzero force acting on it.

EXAMPLE 4.14

Bullet

$m = 0.050$ kg

$D \rightarrow 340$ m/s

Figure 4.28

A bullet of mass 0.050 kg is fired from a gun. It starts from rest and obtains a speed of 340 m/s in a time of 0.0010 seconds. What was the force applied to the bullet?

$$m = 0.050 \text{ kg}$$

$$v_i = 0$$

$$v_f = 340 \text{ m/s}$$

$$t = 0.0010 \text{ s}$$

$$F_{\text{total}} = \frac{p_f - p_i}{t}$$

$$p = m \cdot v$$

$$p_f = (0.050 \text{ kg})(340 \text{ m/s}) = 17 \text{ kg} \frac{\text{m}}{\text{s}}$$

$$p_i = 0$$

$$F_{total} = \frac{p_f - p_i}{t}$$

$$F_{total} = \frac{17\,\frac{kgm}{s} - 0\,\frac{kgm}{s}}{0.0010\,s} = 17{,}000\ N$$

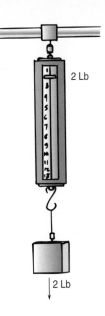

Figure 4.29

Action Reaction

Why does a gun recoil when fired? Why does a rocket move upward when the exhaust gas moves downward? When walking, why do you move forward when you push backward with each step? A law called *action reaction* is responsible.

Newton's third law of motion is sometimes called the law of action reaction. It states that for every force there is an equal and opposite force.

EXAMPLE 4.15

A spring scale weighs a block. The block pulls down with 2 Lb of force. The spring scale pulls up with 2 Lb of force.

EXAMPLE 4.16

A man walking pushes back with 10 Lb of force with each step. He is therefore pushed forward with 10 Lb of force with each step.

Figure 4.30

EXAMPLE 4.17

A rocket exhausts gases downward with 5,000 Lb of thrust (Fig. 4.31). The rocket is therefore pushed upward with 5,000 Lb of force.

Conservation of Momentum

The principle of action reaction is related to a law called the *conservation of momentum,* which states that the total momentum of an isolated system always stays the same.

If we think of a system as a gun and a bullet or a rocket and the rocket's exhaust, the total momentum of these systems stays the same before and after firing. This is described by the equation:

$$p_{total\ before} = p_{total\ after}$$

$$\sum_{before} p_i = \sum_{after} p_i$$

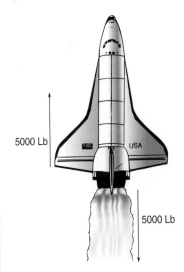

Figure 4.31

This equation says, if you add up the momentum in a system before and after an event, the momentum is the same.

Conservation of momentum can tell us how the speed of an object will change before and after some event takes place. For instance, how fast does a gun recoil when fired?

EXAMPLE 4.18

A gun with a mass of 4.0 kg fires a bullet of mass 0.050 kg with a velocity of 450 m/s. How fast does the gun recoil backward?

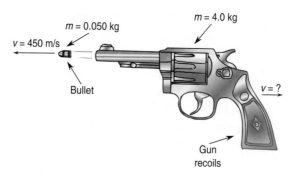

Figure 4.32

$$m_{gun} = 4.0 \text{ kg}$$

$$m_{bullet} = 0.050 \text{ kg}$$

$$v_{bullet} = 450 \frac{m}{s}$$

$$v_{gun} = ?$$

Before the gun is fired the gun and the bullet are still, therefore $P_{total} = 0$. By momentum conservation after the gun is fired, $p_{total} = 0$.

$$p_{gun} + p_{bullet} = 0$$

$$p_{gun} = -p_{bullet}$$

$$m_{gun} \times v_{gun} = -m_{bullet} \times v_{bullet}$$

$$\frac{\cancel{m}_{gun} \times v_{gun}}{\cancel{m}_{gun}} = -\frac{m_{bullet} \times v_{bullet}}{m_{gun}}$$

$$v_{gun} = -\frac{m_{bullet} \times v_{bullet}}{m_{gun}}$$

$$v_{gun} = -\frac{.050 \text{ kg} \times 450 \text{ m/s}}{4.0 \text{ kg}}$$

$$v = -5.6 \frac{m}{s}$$

The gun must move backward to cancel out the momentum of the bullet moving forward.

CHAPTER SUMMARY

Formulas

$$F_w = m \cdot g \quad \leftarrow \quad g = 9.8\,\frac{m}{s^2}\ \text{metric}$$

$$\text{metric or } g = 32.2\,\frac{ft}{s^2}\ \text{English}$$

Force due to weight = mass × acceleration due to gravity

Conversions

1 kg = 1,000 g	1 g is about the mass of a paper clip
1 Lb = 4.45 N	1 newton is almost 1/4 lb
1 slug = 14.6 kg	1 slug has the equivalent weight of 32.2 Lb

Newton's first law says that if a body changes its motion from standing still to moving, or from one speed to another, or changes its direction of travel, it is because there are forces acting upon it that do not balance each other out.

$$\sum_i F_i = m \times a \quad \leftarrow \quad \text{Newton's second law}$$

Sum of Forces = mass × acceleration

Newton's third law of motion is sometimes called the law of action reaction. For every force there is an equal and opposite force.

Static equilibrium: Static equilibrium means all forces are counterbalanced by other forces, and nothing is accelerating. Static equilibrium means that:

$$\sum_i F_i = 0$$

$$p = m \cdot v$$

Momentum = mass × velocity

$$F_f = \mu \cdot F_n$$

Frictional force = coefficient of friction × normal force

$$\mu_s > \mu_k$$

$$F_{total} = \frac{\Delta p}{t} = \frac{p_f - p_i}{t}$$

$$\text{Total Force} = \frac{\text{change in momentum}}{\text{time}}$$

Conservation of momentum means:

$$\sum_{\text{Before}} p_i = \sum_{\text{After}} p_i$$

Problem-Solving Tips

1. Make a sketch of the problem, including a force diagram if relevant.

2. List the quantities you have and the quantity you need.

3. With your list of formulas in front of you, determine the relevant ones. If you have trouble, use the process of elimination.

4. Plug in numbers after you complete all the algebra; sometimes things cancel out.

5. Remember to pay attention to the units. Stay in the correct unit system, mks (meter-kilogram-second) or fps (foot-pound-second).

PROBLEMS

1. A force sensor emits a signal of 10 mV/Lb. Suppose a weight of 100 Lb is added to the scale. What is the voltage output from the sensor?

2. What is the mass of a 180 Lb person in slugs and kilograms? What is the weight of this person in newtons?

3. An automatic door with a mass of 12 kg opens by accelerating at 1 m/s^2. What force must a motor deliver to this door to cause this acceleration?

4. Four tires support a wagon weighing 22 Lb. How much weight does each tire support?

5. Suppose the coefficient of kinetic friction between the rubber tires of a car and a concrete road is $\mu_k = 1.2$. Suppose the car weighs 3,500 Lb. What is the deceleration of the car if it locks up the brakes?

6. What is the momentum of a hammer weighing 0.8 kg striking a nail at 5 m/s?

7. A nail stops the hammer in problem 6 in 0.08 seconds. What force is delivered to the nail?

8. The starship *Enterprise* fires a photon torpedo at a Klingon battle cruiser. Suppose the *Enterprise* has a mass of 1.2×10^{11} kg, and the effective mass of the photon torpedo is 500 kg traveling at 2.9×10^8 m/s. What is the recoil velocity of the *Enterprise?*

9. Suppose a bug hits your car windshield. Describe the force experienced by the bug relative to the force delivered to the windshield. Hint: Think about action reaction.

10. How can the conservation of momentum be used to determine what happened in an automobile accident?

11. Given the set of data shown below from a force sensor determine the equation that models its behavior.

Weight	Resistance
0.50 Lb	1.1 kΩ
1.00 Lb	1.5 kΩ
1.50 Lb	1.9 kΩ
2.00 Lb	2.3 kΩ

12. How would you design a device to measure acceleration (an accelerometer) using a resistive force sensor? Draw a sketch.

13. A bridge must maintain static equilibrium in order for it to keep from collapsing. How could you monitor the bridge to see what kind of load it is under during typical use?

14. In electronics there is a socket for an integrated circuit chip called a zero insertion force (ZIF) socket. This socket is designed to make inserting and removing chips easy. On a 16-pin chip, if 0.25 Lb of force is applied to the chip when inserting, how much force does each pin "feel"?

15. What is the recoil velocity of your computer when it ejects a CD? Assume some reasonable numbers in your calculations for the mass of the computer, disk, and holder and the exit velocity of the CD out of the computer.

Energy, Work, and Power

Energy is used to make things move, light up, and heat up. It takes many forms, and it seems like we are always running out of it. But what exactly is energy? Does a gallon of gasoline have a lot of energy compared with a car battery or what can be obtained from a large solar cell or a windmill? This chapter will attempt to answer these questions, along with explaining how to calculate energy and power. When we put together these ideas of energy and power along with Newton's laws and the equations of motion, we will have a very powerful set of tools for analyzing almost any type of mechanical motion.

Figure 5.1

So, what exactly is energy?
Energy is defined by what it can do. We can use energy to perform work.

Energy: *Energy* is the capacity to do work.

Work: Work in physics is different than work in everyday life. The definition of work is

$$W = F \cdot d$$

$$\text{Work} = \text{force} \times \text{distance}$$

If an object is pushed with a force *(F)* and moves a distance *(d)* in the direction of the applied force, then the work done is *(W)*.

EXAMPLE   5.1

A car is pushed with 55 Lb of force a distance of 6.0 ft. How much work was done?

$$F = 55 \text{ Lb}$$

$$d = 6.0 \text{ ft}$$

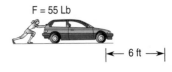

F = 55 Lb

|← 6 ft →|

Figure 5.2

$$W = ?$$

$$W = F \cdot d$$

$$W = (55 \text{ Lb})(6.0 \text{ ft}) = 330 \text{ ft Lb}$$

ENERGY/WORK UNITS

In the English system, work or energy is measured in foot-pounds (ft Lb). In the SI system it is measured in newton-meters (Nm) or joules (J).

English System		SI System
Energy/work	ft Lb	Nm or joules (J)

Conversions:

$1 J = 1 Nm$

$1 J = 0.74 \text{ ft Lb}$

If the applied force is not in the same direction as the motion, however, the work done is:

$W = F \cdot d \cos(\theta)$ ← When the applied force is not in the same direction as the motion, where θ is the angle between the applied force and the motion (Fig. 5.3).

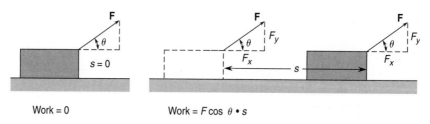

Work = 0   Work = $F \cos \theta \cdot s$

Figure 5.3
Use $W = F \cdot d \cos(\theta)$ when the applied force makes an angle θ with the direction of motion.

E X A M P L E   5 . 2

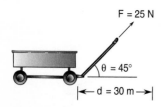

Figure 5.4

A cable is pulling a cart making an angle of 45° with the direction of motion. The tension on the cable is 25 N and pulls the cart 3.0 m. What is the work done?

$$F = 25 \text{ N}$$

$$d = 3.0 \text{ m}$$

$$\theta = 45°$$

$$W = ?$$

$$W = F \cdot d \cos(\theta)$$

$$W = 25 \text{ N} \cdot 3.0 \text{ m} \cdot \cos(45°)$$

$$W = 53 \text{ Nm}$$

$$W = 53 \text{ J} \quad \leftarrow \quad \text{Because } 1 \text{ N} \cdot \text{m} = 1 \text{ J}$$

In a system without any energy loss such as friction, the energy needed to perform some amount of work is equal to the work.

Energy = work performed $\leftarrow$ For a system without any energy loss

EXAMPLE 5.3

How much energy is needed to perform 100 J of work? To perform 100 J of work, 100 J of energy is needed (assuming no energy is lost). Energy and work have the same units.

 ## EFFICIENCY

Efficiency is a measure of how much energy is converted into doing some useful work. For a system that is not friction free, the efficiency will always be less than 100%.

$$\text{Efficiency} = \left(\frac{\text{work done}}{\text{energy consumed}} \right) \times 100$$

EXAMPLE 5.4

An automobile engine performs 9,500 J of work in order to move a car. The car consumes 47,500 J of energy by burning gasoline. What is the efficiency of this car?

$$W = 9,500 \text{ J}$$

$$E = 47,500 \text{ J}$$

$$\text{Efficiency} = ?$$

$$\text{Efficiency} = \left(\frac{\text{work done}}{\text{energy consumed}} \right) \times 100$$

$$\text{Efficiency} = \left(\frac{9,500 \text{ J}}{47,500 \text{ J}} \right) \times 100$$

$$\text{Efficiency} = 20 \%$$

Figure 5.5

Energy can take two basic forms. *Potential energy* is the energy stored in a system due to the arrangement of the position of its parts. A compressed spring, a weight raised to some height, and a battery with its charges separated all have potential energy (Fig. 5.6). This energy is stored, waiting to be used.

Kinetic energy means moving energy. A hammer striking the head of a nail delivers its kinetic energy to the nail, forcing it into the wood some distance, performing some work (Fig. 5.7).

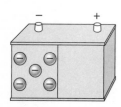

Figure 5.7
A hammer delivers its kinetic energy to a nail.

Figure 5.6
A weight raised to some height, a compressed spring, and a battery with its charges separated all have potential energy.

STORED OR POTENTIAL ENERGY SOURCES

Gravitational

Energy can be stored by lifting something above the ground. This is *gravitational potential energy*. The energy can then be released, allowing it to fall back to the ground, turning this potential into kinetic energy (Fig. 5.8).

Figure 5.8
A pile driver. Potential energy is stored, then released, and turned into kinetic energy. The kinetic energy is delivered to the pile, performing some work.

Gravitational potential energy is given by the following formula.

$$PE = m \cdot g \cdot h$$

Potential energy = mass × gravity × height

EXAMPLE 5.5

A water tower stores potential energy in the water by pumping it up into a tank, high above the ground. Suppose the tank is 10.0 m above the ground and contains 12,500 kg of water. What is the potential energy of this water?

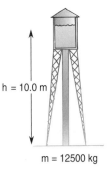

$h = 10.0$ m

$$h = 10.0 \text{ m}$$
$$m = 12{,}500 \text{ kg}$$
$$PE = ?$$
$$PE = m \cdot g \cdot h$$
$$PE = (12{,}500 \text{ kg})(9.8 \, \frac{\text{m}}{\text{s}^2})(10.0 \text{ m})$$
$$PE = 1.2 \times 10^6 \text{ J}$$

$m = 12500$ kg

Figure 5.9

Springs

In order to compress or elongate a spring, work must be done on the spring. The energy used to perform this work is stored in the spring, waiting to be released (Fig. 5.10).

Batteries

A *battery* is a source of potential energy (Fig. 5.11).
     Every battery has two values associated with it, its voltage and its amp-hour rating. The *amp-hour rating* is a measure of how much charge is stored in the battery. The larger the amp-hour rating, the longer the battery will last.

Figure 5.10
A compressed or stretched spring stores potential energy.

$$Ah = A \times h$$

Amp-hour = Amps × hours

An amp-hour rating of 100 means the battery can put out

$$100 \text{ amp-hour} \quad \rightarrow \quad \begin{array}{l} 100 \text{ amps for 1 hour} \\ 50 \text{ amps for 2 hours} \\ 10 \text{ amps for 10 hours} \end{array}$$

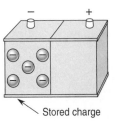

Stored charge

Figure 5.11
A battery separates charge to be used later to perform some useful work.

The greater the current drawn out of the battery, the shorter the battery's life. See Table 5.1 for various battery amp-hr ratings.

Table 5.1
Characteristics of Some Common Batteries

Battery	Voltage	Ampere-hour rating	Mass, g
Primary			
"pen-light"	1.5	0.58	20
D-cell	1.5	3.0	90
#6 dry cell	1.5	30	860
#2U6 battery	9.0	0.325	30
#V-60 battery	90	0.47	580
#1 mercury cell	1.4	1.0	12
#42 mercury cell	1.4	14	166
Secondary (rechargeable)			
#CD-21 nickel-cadmium	6.0	0.15	52
#CD-29 nickel-cadmium	12.0	0.45	340
lead-acid car battery	6.0	84	13 000
lead-acid car battery	12.0	96	25 000

EXAMPLE 5.6

Suppose you leave your car's lights on in a parking lot. If the car battery has 100 Ah of charge and the headlights pull 12 A each, how long before the battery is dead?

$$\text{amp} \times \text{h} = 100 \text{ Ah}$$

$$I = 12 \text{ A}$$

$$I_T = 12 \text{ A} + 12 \text{ A} = 24 \text{ A} \quad \leftarrow \quad \text{Two headlights worth of current}$$

$$\text{h} = ?$$

$$\text{Ah} = \text{A} \times \text{h}$$

$$\frac{\text{Ah}}{\text{A}} = \frac{\cancel{\text{A}} \times \text{h}}{\cancel{\text{A}}} \quad \leftarrow \quad \text{Solving for hours}$$

$$\text{h} = \frac{\text{Ah}}{\text{A}}$$

$$\text{h} = \frac{100 \text{ Ah}}{24 \text{ A}} = 4.2 \text{ h} \quad \leftarrow \quad \text{Battery will be dead in this time}$$

As stated above, the amp-hour rating is a measure of how much charge is stored in the battery. To see how to convert from amp-hours into the unit of charge, let's review the units of charge and current.

ELECTRICAL UNITS

Charge (Q) is the unit of charge measured in coulombs. Amps is a measure of how much charge flows past a point in some time.

$$I \text{ (amps)} = \frac{Q \text{ (coulombs)}}{t \text{ (seconds)}}$$

Therefore

$$Q = I \times t$$

To convert an amp-hour value to charge, we must convert the hour into seconds (1 h = 3,600 s). For example, convert 100 Ah into coulombs.

$$100 \text{ A}\cancel{h}\left(\frac{3,600 \text{ s}}{1 \cancel{h}}\right) = 360,000 \text{ coulombs}$$

The amount of energy a battery can store is determined by how much charge it can store and its voltage:

$$E = Q \times V$$
$$\text{Energy} = \text{charge} \times \text{voltage}$$

The units these three quantities are measured in are:

$$E\text{(joules)} = Q\text{(coulombs)} \times V\text{(volts)}$$

EXAMPLE 5.7

Suppose a 12-V car battery has a 90-Ah rating. How much energy is it storing when it is fully charged?

$$Ah = 95 \text{ Ah}$$
$$V = 12 \text{ V}$$
$$E = ?$$
$$E = Q \times V$$
$$Q = 95 \text{ A}\cancel{h}\left(\frac{3,600 \text{ s}}{1 \cancel{h}}\right)$$
$$Q = 340,000 \text{ coulombs}$$
$$E = Q \times V$$
$$E = 340,000 \text{ C} \times 12 \text{ V}$$
$$E = 4,100,000 \text{ J or } 4.1 \text{ MJ}$$

EXAMPLE 5.8

Suppose an electric car is powered by a 12-V battery containing 4.1 MJ of energy. If it takes 150 Lb or 668 N of force to move the car at a constant rate, how far can the car travel? Assume an efficiency of 70%.

$$E = 4.1 \text{ MJ}$$
$$F = 668 \text{ N}$$

$$\text{Efficiency} = 70\%$$

$$d = ?$$

Since there is only 70% efficiency, the total energy available to do work is

$$E = 0.70 \times 4.1 \text{ MJ}$$

$$E = 2.87 \text{ MJ}$$

$$E = F \cdot d$$

$$\frac{E}{F} = \frac{\cancel{F} \cdot d}{\cancel{F}}$$

$$d = \frac{E}{F}$$

$$d = \frac{2.87 \times 10^6 \text{ J}}{668 \text{ N}}$$

$$d = 4{,}296 \text{ m}$$

This is only about 3 miles! Better increase the number of batteries!

Fuels

Figure 5.12
Chemical potential energy is analogous to the energy stored in compressed springs.

Fuels store chemical potential energy. A molecule of gasoline can be thought of as a bunch of atoms separated by compressed springs (Fig. 5.12). The springs are the intermolecular force between atoms. With the addition of heat, the springs will be released from compression, hurling the atoms outward. This is an explosion of the gas molecule that turns the chemical potential into kinetic energy. Listed below in Table 5.2 is the energy released when these fuels are burned.

Table 5.2
Energy Contained in Fuels

Fuel	Heat of Combustion		Air-fuel Ratio for Complete Combustion, lb_m/lb_m or g/g
	Btu/lb_m	MJ/kg	
Butane	20 000	46	15.4
Gasoline	19 000	44	14.8
Crude oil	18 000	42	14.2
Fuel oil	17 500	41	13.8
Coke	14 500	34	11.3
Coal, bituminous	13 500	31	10.4

Actual values will vary slightly depending on geographical origin. Values do not include latent heat in any water vapor formed as a product of combustion.
Note: 1 MJ/kg = 1000 kJ/kg = 1000 J/g

EXAMPLE 5.9

Calculate the gas mileage a car would get if the force needed to propel it along at some speed is 215 Lb and the efficiency of the car is 20%.

$$F = 215 \text{ Lb}$$

$$\text{Efficiency} = 20\%$$

The energy needed to travel 1 mile, or 5,280 ft, is:

$$W = F \times d$$

$$W = 215 \text{ Lb} \times 5,280 \text{ ft}$$

$$W = 1.14 \times 10^6 \text{ ft Lb}$$

The car will expend more energy than this because it is only 20% efficient.

The energy consumed traveling 1 mile can be determined from:

$$\text{Efficiency} = \left(\frac{\text{work done}}{\text{energy consumed}} \right) \times 100$$

$$\text{Energy consumed} = \left(\frac{\text{work done}}{\text{efficiency}} \right) \times 100$$

$$\text{Energy consumed} = \left(\frac{1.14 \times 10^6 \text{ ft Lb}}{20} \right) \times 100$$

$$\text{Energy consumed} = 5.70 \times 10^6 \text{ ft Lb}$$

Since the calculation was based on traveling 1 mile, the energy consumed per mile is:

$$\text{Energy consumed/mile} = 5.70 \times 10^6 \frac{\text{ft Lb}}{\text{mile}}$$

Gasoline has an energy content of 9.7×10^7 ft Lb/gal. The gas mileage is therefore:

$$\text{Gas mileage} = \frac{9.7 \times 10^7 \text{ ft Lb/gal}}{5.70 \times 10^6 \dfrac{\text{ft Lb}}{\text{mile}}}$$

$$\text{Gas mileage} = 17 \frac{\text{miles}}{\text{gallon}}$$

Can you estimate the gas mileage of your car from a calculation like this?

KINETIC ENERGY

Kinetic energy means moving energy. If something moving strikes something stationary, it will deliver a force to it, causing it to move and imparting its energy to the stationary object. The formula for kinetic energy is

$$\text{KE} = \frac{1}{2} m \cdot v^2$$

$$\text{Kinetic energy} = \frac{1}{2} \text{ mass} \times \text{velocity}^2$$

EXAMPLE 5.10

A rail gun consists of a U-shaped conductor with a sliding conducting rail lying across it. When a pulse of current is induced into the circuit, the rail will quickly move outward due to the magnetic fields generated in the loop.[1] Suppose a marble with a mass of 0.060 kg is placed on the rail and launched at a speed of 450 m/s. What is the kinetic energy of the marble?

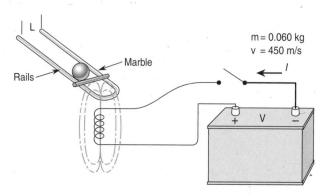

Figure 5.13

$$m = 0.060 \text{ kg}$$

$$v = 450 \, \frac{\text{m}}{\text{s}}$$

$$KE = \frac{1}{2} mv^2$$

$$KE = \frac{1}{2} (0.060 \text{ kg}) \left(450 \, \frac{\text{m}}{\text{s}} \right)^2$$

$$KE = 6{,}100 \text{ J}$$

EXAMPLE 5.11

How much work can be performed by an object that has 6,100 J of kinetic energy? If there is a 100% efficient transfer of energy into work, then 6,100 J of energy can perform 6,100 J of work. If the energy transfer into work is less than 100%, say 40%, the amount of work that can be done is:

$$40\% \text{ of } 6{,}100 \text{ J} = 0.4 \times 6{,}100 \text{ J} = 2{,}400 \text{ J}$$

EXAMPLE 5.12

A batter hits a baseball with a mass of 2.1 kg delivering 2,500 J of energy to it. At what speed is the ball traveling?

Figure 5.14

---

[1] A magnetic field is generated around a current-carrying wire. When two current carrying wires are parallel to each other and the current in each wire is moving in the opposite direction, the wires will repel each other. This is similar to the repulsive force between two magnets, with like poles facing each other.

$$m = 2.1 \text{ kg}$$

$$KE = 2{,}500 \text{ J}$$

$$KE = \frac{1}{2}mv^2$$

$$2KE = 2\left(\frac{1}{2}mv^2\right) \quad \leftarrow \quad \text{Solving for } v$$

$$2KE = mv^2$$

$$\frac{2KE}{m} = \frac{mv^2}{m}$$

$$v^2 = \frac{2KE}{m}$$

$$v = \sqrt{\frac{2KE}{m}}$$

$$v = \sqrt{\frac{2 \cdot 2500 \text{ J}}{2.1 \text{ kg}}} = 49 \, \frac{\text{m}}{\text{s}}$$

EXAMPLE 5.13

A hammer with 500 J of kinetic energy strikes a nail, driving it 0.01 m into a piece of wood. With what average force did the hammer strike the nail?

Figure 5.15

$$KE = 500 \text{ J}$$

$$d = 0.01 \text{ m}$$

$$F = ?$$

$$\text{work} = \text{energy}$$

$$\text{work} = 500 \text{ J}$$

$$w = F \cdot d$$

$$\frac{w}{d} = \frac{F \cdot d}{d} \quad \leftarrow \quad \text{solving for } F$$

$$F = \frac{w}{d}$$

$$F = \frac{500 \text{ J}}{0.01 \text{ m}} = 50{,}000 \text{ N}$$

 ## ENERGY CONSERVATION

The total energy of an isolated system is a constant. Energy is never created or destroyed; it just changes form. Even for a system that is less than 100% efficient, we can account for all of the energy used if we add up the work done, the energy wasted due to friction that went into heating up the system, any light or sound given off, or any other form the input energy may have turned into. Given this law,

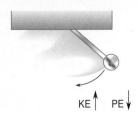

if something falls, losing some of its potential energy, the loss must be made up by gaining kinetic energy (neglecting air resistance).[2] Kinetic and potential energy are in balance. A pendulum illustrates this nicely (Fig. 5.16).

Figure 5.16
Consider a pendulum. If the potential energy decreases, the kinetic energy increases.

EXAMPLE 5.14

The water at the top of a waterfall has potential energy. If this water is allowed to fall back to the ground, its potential energy will be turned into kinetic energy (Fig. 5.17). Suppose the potential energy of the water is 54,000 J at a height of 10.0 m. What is the kinetic energy of this water when it falls to the ground, and what is its velocity?

$$PE = 54,000 \text{ J}$$
$$h = 10.0 \text{ m}$$
$$v = ?$$
$$KE = ?$$

After falling, all this potential energy will be converted into kinetic energy.

$$KE = 54,000 \text{ J}$$
$$KE = PE$$
$$\frac{1}{2}mv^2 = mgh$$
$$\frac{1}{2}\cancel{m}v^2 = \cancel{m}gh$$
$$\frac{1}{2}v^2 = gh$$
$$\cancel{2}\frac{1}{\cancel{2}}v^2 = 2gh$$
$$v^2 = 2gh$$
$$v = \sqrt{2\,gh}$$
$$v = \sqrt{2\left(9.8\,\frac{\text{m}}{\text{s}^\text{s}}\right)(10.0 \text{ m})}$$
$$v = 14\,\frac{\text{m}}{\text{s}}$$

Figure 5.17
The potential energy stored is turned into kinetic energy when the water flows downward.

POWER

Power is the rate at which energy is used.

---

[2] We are, of course, neglecting air resistance, which will limit the object's velocity to a terminal velocity.

$$P = \frac{E}{t}$$

$$\text{Power} = \frac{\text{energy}}{\text{time}}$$

If a lot of energy is used quickly, then a lot of power is consumed.

Power is measured in watts, $\frac{\text{ft} \cdot \text{lb}}{\text{s}}$ or horsepower.

English units	**Metric units**
Power $\frac{\text{ft} \cdot \text{lb}}{\text{s}}$ or horse power (hp)	watts (W)

Conversions:

$$1 \text{ hp} = 550 \frac{\text{ft} \cdot \text{lb}}{\text{s}}$$

$$1 \text{ hp} = 746 \text{ W}$$

$$1 \text{ W} = 0.74 \frac{\text{ft} \cdot \text{lb}}{\text{s}}$$

EXAMPLE 5.15

A light bulb consumes 100 J of energy every 2 seconds. How much power is being consumed?

$$E = 100 \text{ J}$$
$$T = 2 \text{ s}$$
$$P = \frac{E}{t}$$
$$P = \frac{100 \text{ J}}{2 \text{ s}}$$
$$P = 50 \text{ W}$$

Figure 5.18

EXAMPLE 5.16

A crane lifts a crate with a force of 250 lb, $1\overline{0}$ ft in the air in 5 seconds. How much work was done? How much energy was used and power consumed?

$$F = 250 \text{ Lb}$$
$$d = 1\overline{0} \text{ ft}$$
$$t = 5 \text{ s}$$
$$W = f \cdot d$$

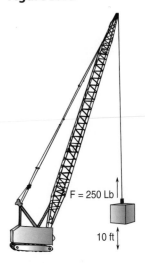

F = 250 Lb

10 ft

Figure 5.19

$$W = (250 \text{ Lb})(1\overline{0} \text{ ft}) = 2{,}500 \text{ ft} \cdot \text{Lb}$$

$$\text{Energy} = \text{work}$$

$$\text{Energy} = 2{,}500 \text{ ft} \cdot \text{Lb}$$

$$P = \frac{E}{t}$$

$$P = \frac{2{,}500 \text{ ft Lb}}{5 \text{ s}}$$

$$P = 500 \frac{\text{ft} \cdot \text{Lb}}{\text{s}}$$

EXAMPLE 5.17

Figure 5.20

A lawn mower is rated at 3 hp. What is this power in watts?

$$P = 3.0 \text{ hp}$$

$$1 \text{ hp} = 746 \text{ W}$$

$$3 \text{ hp} \left(\frac{746 \text{ W}}{1 \text{ hp}} \right) = 2{,}200 \text{ W}$$

We can put together the power formula $P = \dfrac{E}{t}$ with the formula for electrical power $P = I \cdot V$ to get:

$$E = I \cdot V \cdot t$$

$$\text{Energy} = \text{current} \times \text{voltage} \times \text{time}$$

EXAMPLE 5.18

A motorized cart is powered by a 12-V battery, and the motor is pulling 15 A. a) If we assume the motor is 100% efficient, how much energy is generated in 10.0 seconds? b) If the motor delivers 5.0 N of force to the wheels, how far did the cart move in this 10.0 seconds? We will assume there is no friction.

a)
$$V = 12 \text{ V}$$

$$I = 15 \text{A}$$

$$t = 10.0 \text{ s}$$

$$F = 5.0 \text{ N}$$

Assume 100% efficiency and no friction.

$$E = ?$$

$$E = I \cdot V \cdot t$$

$$E = 15 \text{A} \cdot 12 \text{ V} \cdot 10.0 \text{ s}$$

$$E = 1{,}800 \text{ J}$$

b)

$$F = 5.0 \text{ N}$$

$$W = F \cdot d$$

$$E = W \quad \leftarrow \quad \text{For a 100\% efficient machine}$$

$$E = F \cdot d$$

Solving for d gives

$$d = \frac{E}{F}$$

$$d = \frac{1,800 \text{ J}}{5.0 \text{ N}} = 360 \text{ m}$$

KILOWATT-HOURS

The power meter on your house keeps track of how much energy you use (Fig. 5.21). It does this by keeping track of how many kilowatts are used and the length of time those kilowatts are used (remember energy = power × time). To determine the power used, the meter simultaneously measures the voltage and the current drawn. The meter reading is given in kilowatt-hours (kWh), because we pay for energy, not power.

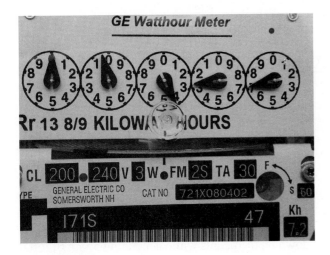

Figure 5.21
Electrical power meter.

Kilowatt-hours = kilowatts × hours

P = 1.2 kW
t = 0.1 hr

Figure 5.22

EXAMPLE 5.19

A hair dryer consumes 1.2 kW of power and is run for 0.1 hours. What is the energy consumed?

$$P = 1.2 \text{ kW}$$
$$t = 0.1 \text{ h}$$
$$\text{kWh} = \text{kW} \times \text{h}$$
$$\text{kWh} = (1.2 \text{ kW})(0.1 \text{ h}) = 0.12 \text{ kWh}$$

EXAMPLE 5.20

The electrical power company charges you by the number of kilowatt-hours used. Suppose the power company charges 9 cents/kWh. How much does it cost you to leave a 100-W light on for 8 hours? This problem is, effectively, a conversion from kilowatt-hours to cents.

$$P = 100 \text{ W}$$
$$t = 8 \text{ h}$$
$$\text{Rate} = 9 \text{ cents/kWh}$$
$$\text{Cost} = ?$$
$$100 \text{ W} = 0.1 \text{ kW}$$
$$\text{kWh} = \text{kW} \times \text{h}$$
$$\text{kWh} = 0.1 \text{ kW} \times 8 \text{ h}$$
$$\text{kWh} = 0.8 \text{ kWh}$$
$$0.8 \text{ kWh} \left(\frac{9 \text{ cents}}{1 \text{ kWh}} \right) = 7.2 \text{ cents}$$

POWER, FORCE, AND VELOCITY

Power, force, and velocity are related through the formula:

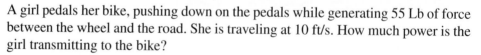

$$P = F \cdot v$$
$$\text{Power} = \text{force} \times \text{velocity}$$

EXAMPLE 5.21

$v = 10$ ft/s $F = 55$ Lb

Figure 5.23

A girl pedals her bike, pushing down on the pedals while generating 55 Lb of force between the wheel and the road. She is traveling at 10 ft/s. How much power is the girl transmitting to the bike?

$$F = 55 \text{ Lb}$$
$$v = 10 \text{ ft/s}$$
$$P = ?$$
$$P = F \cdot v$$
$$P = (55 \text{ Lb}) \cdot (10 \text{ ft/s})$$
$$P = 550 \text{ ft Lb/s}$$
$$P = 1 \text{ hp}$$

EXAMPLE 5.22

You are asked to design an elevator to transport 1,900 Lb of freight at a constant speed using a 10-hp motor. How fast will the elevator car move?

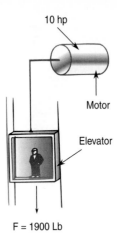

$$F = 1,900 \text{ Lb}$$

$$P = 10 \text{ hp} \left(\frac{550 \text{ ft lb/s}}{1 \text{ hp}} \right) = 5,500 \text{ ft Lb/s}$$

$$v = ?$$

$$P = F \cdot v$$

$$\frac{P}{F} = \frac{F \cdot v}{F} \quad \leftarrow \quad \text{Solving for } v$$

$$v = \frac{P}{F}$$

$$v = \frac{5,500 \text{ ft Lb/s}}{1,900 \text{ Lb}}$$

$$v = 2.9 \text{ ft/s}$$

Figure 5.24

OTHER SOURCES OF ENERGY

Fuel Cells

Fuel cells function like a battery in that they derive their electrical energy from a chemical reaction (Fig. 5.25). Unlike a battery, they never need to be recharged. They do require fuel, however, in the form of hydrogen and oxygen. A fuel cell consists of two electrodes contained in an electrolyte. Hydrogen flows over one electrode, and oxygen flows over the other. A chemical reaction takes place at the electrodes, removing an electron from the hydrogen, attaching itself to one electrode (the cathode) and the positive proton to the other electrode (the anode). This buildup of opposite charge on two electrodes makes for a device that functions like a battery. For more information on types of fuel cells and how to build your own, look on the web at http://216.51.18.233/index_e.html.

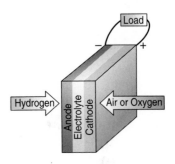

Figure 5.25
A fuel cell.

Wind

A *windmill* can make the kinetic energy of the wind pump water, mill corn, generate electricity, and so on (Fig. 5.26). How much energy does wind have? The energy contained in the wind depends on its velocity, density, and moisture content.

Hydroelectric Energy

Hydroelectric energy is the energy captured from falling water (Fig. 5.27).

Question: What is the difference between gas, coal, and nuclear power? The electrical energy generated from gas, coal, and nuclear fuel is not a direct conversion. The heat generated from these three fuel sources is used to boil water, to make steam. This steam is then directed at a turbine to turn an electrical generator (Fig. 5.28).

Figure 5.26
A windmill converts the wind's kinetic energy into rotational energy to be used for some purpose, such as generating electricity or pumping water.

Figure 5.27
Hydroelectric power is power generated from falling water.

Solar Energy

Solar cells convert the energy contained in sunlight into electrical energy (Fig. 5.29). On a sunny day about 1,000 W/m^2 of energy is contained in the sunlight that reaches earth. Currently, solar cells can at best convert about 20% of this energy into electricity. A solar cell is basically a very thin diode, a PN junction, with one side transparent to light. As light makes its way to the junction, it knocks free electrons on the P side, and they are transported to the N side because of the electric field present in every PN junction. A grid is placed over the solar cell to collect these electrons on the front, and a metal film on the back acts as the positive terminal. The cell now acts like a battery, deriving its energy from light instead of some chemical reaction (Fig. 5.30).

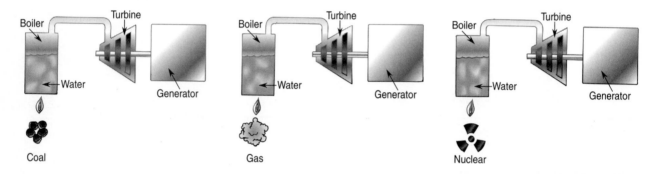

Figure 5.28
Gas, coal, and nuclear power are all used to heat water to generate steam, to turn a turbine connected to an electrical generator. The only difference between them is the fuel used to heat the water.

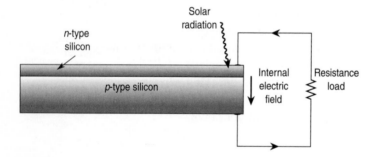

Figure 5.29
A solar cell.

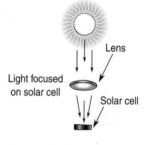

Figure 5.30
Lenses help collect more light to focus on the solar cell.

Currently, at about $0.5 to $1 per kWh, solar power is much more expensive than electricity from the utility companies (about $0.09 per kWh). Only in areas where there is a lot of sunlight and where there is no electric utility has it become economically viable, at this time.

Figure 5.31
A lightning bolt contains millions of joules of energy.

Lightning

A bolt of *lightning* contains millions of joules of energy (Fig. 5.31). It would be great if we could capture and use it. Research is currently seeking ways to capture this energy. One of the biggest problems to overcome in this field is finding a way to collect the energy without destroying the equipment used to collect it!

CHAPTER SUMMARY

Energy is the capacity to do work.

$$W = F \cdot d$$

$$W = F \cdot d \cos(\theta)$$

$$\text{Work} = \text{force} \times \text{distance}$$

Conversions:

$$1 \text{ J} = 1 \text{ Nm}$$

$$1 \text{ J} = 0.74 \text{ ft Lb}$$

$$1 \text{ hp} = 550 \frac{\text{ft} \cdot \text{Lb}}{\text{s}}$$

$$1 \text{ hp} = 746 \text{ W}$$

$$1 \text{ W} = 0.74 \frac{\text{ft} \cdot \text{Lb}}{\text{s}}$$

$$\text{Efficiency} = \left(\frac{\text{work done}}{\text{energy consumed}} \right) \times 100$$

$$\text{PE} = m \cdot g \cdot h$$

$$\text{Potential energy} = \text{mass} \times \text{gravity} \times \text{height}$$

$$\text{KE} = \frac{1}{2} m \cdot v^2$$

$$\text{Kinetic energy} = \frac{1}{2} \text{mass} \times \text{velocity}^2$$

$$\text{Amp-hours} = \text{Amps} \times \text{hours}$$

$$I = \frac{Q}{t}$$

$$\text{Current(amps)} = \frac{\text{charge (coulombs)}}{\text{time (seconds)}}$$

$$P = \frac{E}{t}$$

$$\text{Power} = \frac{\text{energy}}{\text{time}}$$

$$P = F \times v$$

$$\text{Power} = \text{force} \times \text{velocity}$$

$$E = I \cdot V \cdot t$$

$$\text{Energy} = \text{current} \times \text{voltage} \times \text{time}$$

$$\text{kWh} = \text{kilowatts} \times \text{hours}$$

Problem-Solving Tips

- All problems dealing with electricity and energy should be done using the mks system of units.
- There are a lot of formulas in this chapter, so keep a formula sheet in front of you while solving the problems.

PROBLEMS

1. A computer's CD drive pushes out a CD holder with 0.5 Lb of force, through a distance of 0.4 ft, and takes only 1 second to eject. How much work was done? How much energy was consumed?

2. Convert 25 J into foot-pounds.

3. An elevator motor consumes 81,000 J of energy lifting 3,500 N of weight at a constant velocity a distance of 4 m. What is the efficiency of this motor?

4. How much energy is required to pick a 1-kg book off the ground and raise it to a height of 1.2 m?

5. A battery having an amp-hour rating of 20 Ah is connected to a light bulb pulling 0.5 A. How long will the light remain lit?

6. The boxer Mike Tyson delivers a punch to his opponent. Suppose Tyson's fist weighs 5 Lb and is moving at 10 ft/s. What is the energy contained in his fist? (Remember to first convert the weight to mass.) If all this energy is delivered to his opponent, pushing his jaw back 1 in, what is the average force contained in his punch? (Remember to convert inches to feet.)

7. You build an electric generator by connecting your 3.5-hp lawn mower engine to a car

alternator. Assume the efficiency of this system is 40%. How many watts can you generate with this generator?

8. What is the minimum power required of a conveyer belt to lift a 150-kg package a height of 2.0 m in 3.0 seconds?

9. In a hydroelectric power plant, suppose that water falls a distance of 110 m before reaching the turbine/generator. What is the speed of the water when it reaches the turbine?

10. Suppose the water in problem 9 strikes the turbine with 1.2×10^6 N of force. What power is delivered to the turbine? Suppose the efficiency of this system is 21%. How many watts does the system put out?

11. How much work is done in pulling a wagon with a force of 15 Lb a distance of 10.0 ft when the handle makes an angle of 45° with the horizontal?

12. Use energy conservation to calculate how fast an apple will hit Isaac Newton's head if he is sitting below the tree and the apple falls 5.0 ft onto his head.

13. What is the approximate amount of energy a bat must deliver to a baseball if it is to knock it out of the park a distance of 450 ft?

14. An electric car is driven by a 65-hp electric motor that is 80% efficient. If it is traveling at 55 mph, for a distance of 56 miles, how much energy did the batteries consume?

15. If the batteries in the car in problem 14 have 95 Ah of charge each, how many batteries are required to travel the 56 miles?

Rotational Motion

Figure 6.1

In this chapter we will discuss rotational motion. We will apply our equations to wheels, motors, and anything else that rotates.

Any point on a circle can be described by its radius r from the center and an angle θ. The angle can be measured in degrees, revolutions, or radians (Fig. 6.2). Comparing the three units:

$$1 \text{ revolution} = 360° = 2\pi \text{ radians}$$

See Figure 6.3 for a comparison of angular units.

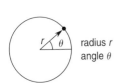

Figure 6.2
A point on a circle can be described by its radius r from the center and an angle θ measured in degrees, revolutions, or radians.

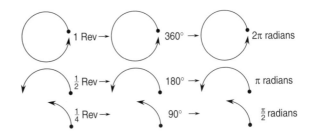

Figure 6.3
A comparison of angular units.

The unit *radians* may be unfamiliar. It has its origins in nature. Take any circle and place a string around it to measure its circumference. Then measure its diameter. Divide the circumference by the diameter, and the number 3.14159…, which is written as π, will result.

Converting Between Angular Units
$$1 \text{ rev} = 360° = 2\pi \text{ rad}$$

E X A M P L E   6 . 1

How many revolutions are there in 180°?

$$\theta = 180°$$
$$1 \text{ rev} = 360°$$
$$180° \left(\frac{1 \text{ rev}}{360°} \right) = 0.50 \text{ rev}$$

EXAMPLE 6.2

How many radians are there in 320°?

$$\theta = 320°$$
$$2\pi \text{ rad} = 360°$$
$$320° \left(\frac{2\pi \text{ rad}}{360°} \right) = 5.6 \text{ rad}$$

EXAMPLE 6.3

How many radians are there in 200 revolutions?

$$\theta = 200 \text{ rev}$$
$$1 \text{ rev} = 2\pi \text{ rad}$$
$$200 \text{ rev} \left(\frac{2\pi \text{ rad}}{1 \text{ rev}} \right) = 1,260 \text{ rad}$$

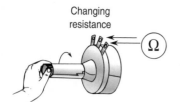

Figure 6.4
A potentiometer used to measure angular orientation.

MEASURING ANGLES

Aside from using a protractor, there are a number of electronic ways of measuring an angle.

Potentiometer

A potentiometer connected to the center of rotation can be used to determine the angular position of something undergoing a rotation (Fig. 6.4).

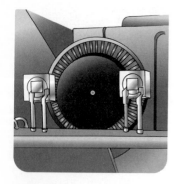

Figure 6.5
A computer mouse converts the revolutions of the ball using an optical encoder in the mouse to a position on the computer screen.

Optical Rotary Encoder

A computer mouse contains two small wheels with spokes that are driven by the ball at the bottom of the mouse. A light source and detector are placed on either side of the wheel. The light detector measures how many times the light beam is broken as the wheel turns. Simply counting the number of interruptions of the light beam determines the angle through which the ball rotated or the distance the mouse has moved (Fig. 6.5).

Gyroscope

A gyroscope consists of a spinning wheel mounted sideways on a frame. Because the wheel is spinning, its momentum resists having its orientation changed. As the instrument is rotated, the spinning wheel will remain pointing in the same direction. The angle measured is the angle between the instrument and the spinning wheel with the fixed orientation (Fig. 6.6).

DISTANCE MEASUREMENTS ALONG A CIRCLE

Any motion that takes place in the circle can be described with two parameters:

1. The distance (r) the object is away from the center of rotation
2. The angle (θ) through which it moves

Suppose a point travels along the arc of a circle. How far has this point traveled?

$$d = r \cdot \theta \quad \leftarrow \quad \theta \text{ must be in radians}$$
Distance along an arc = radius × angle (in radians)

If a wheel of radius r rolls along a surface through an angle θ, the wheel has traveled a distance d (Fig. 6.7):

$$d = r \cdot \theta \quad \leftarrow \quad \theta \text{ must be in radians}$$
Distance a wheel rolls = radius × angle (in radians)

EXAMPLE 6.4

A point on the rim of a wheel with a radius of 2.0 ft travels through an angle of 1.5 radians (Fig. 6.8). a) How far did the point travel on the wheel? b) If the wheel was rolling on the ground, how far did it roll?

a)

$r = 2.0$ ft

$\theta = 1.5$ rad

$d = ?$

$d = r \cdot \theta$

$d = (2.0 \text{ ft})(1.5 \text{ rad})$

$d = 3.0$ ft $\quad \leftarrow \quad$ We don't retain the unit rad.

b)
The distance the wheel rolls is the same as the distance the point travels through on the rim, $d = 3.0$ ft

Figure 6.6
A gyroscope demonstrated using a spinning bicycle wheel hanging by a cable.

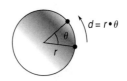

Figure 6.7
If a wheel of radius r rolls along a surface through an angle θ, the wheel has traveled a distance d, $d = r \cdot \theta$, so θ must be in radians.

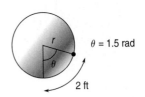

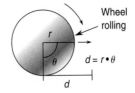

Figure 6.8

Note the unit radian drops out in the calculation of $d = r \cdot \theta$ because the unit radian is dimensionless.

When using the formula $d = r \cdot \theta$, if θ were in any unit other than radians, the answer would be wrong. Why are there so many different units for angles, and when do you use which one? The answer depends on the context of the problem.

As a general rule, however, always use radians when a calculation involves an angle times another quantity.

EXAMPLE 6.5

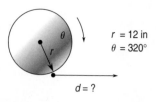

Figure 6.9

A wheel with a radius of 12 in rotates through an angle of 320°. How far has this wheel traveled?

$$r = 12 \text{ in}$$

$$\theta = 320°$$

$$d = ?$$

$d = r\theta$ In order to use this fomula, θ must be in radians.

$$(320°)\left(\frac{2\pi}{360°}\right) = 5.6 \text{ rad}$$

$$d = (12 \text{ in})(5.6 \text{ rad})$$

$$d = 67 \text{ in}$$

EXAMPLE 6.6

Figure 6.10

Suppose a wheel with a radius of 1.3 m rolls along the ground through an angle of 1.5 revolutions. How far has the wheel traveled?

$$r = 1.3 \text{ m}$$

$$\theta = 1.5$$

$$d = ?$$

$d = r\theta$ In order to use this fomula, θ must be in radians.

$$(1.5 \text{ rev})\left(\frac{2\pi}{1 \text{ rev}}\right) = 9.4 \text{ rad}$$

$$d = (1.3 \text{ m})(9.4 \text{ rad})$$

$$d = 12 \text{ m}$$

■ ANGULAR VELOCITY

Figure 6.11
Angular velocity is a measure of how fast something is spinning.

Angular velocity is the measure of how fast something is spinning (Fig. 6.11). Everyone is familiar with the unit revolutions per minute (rpm), which is a measure of how fast something is spinning. It describes how many revolutions a shaft or a wheel turns in 1 minute.

EXAMPLE 6.7

A shaft on a motor turns 6,000 revolutions in 1 minute. What is its angular velocity?

$$6{,}000 \text{ rev/1 min} = 6{,}000 \text{ rpm}$$

$$\text{rpm} = \text{rev/min}$$

Measuring Angular Velocity

If a shaft or wheel is moving slowly enough, simply counting the number of revolutions in a given amount of time is enough to determine the angular velocity. However, it's more probable the wheel is turning too fast to perform such a measurement, and therefore another means is necessary.

Tachometer

A tachometer is used to determine angular velocity. There are a number of types of tachometers. Typically they are simply proximity sensors such as the one described below.

Magnetic Pickup Coil Tachometer

In Figure 6.12, a coil is placed near the rotating gear. The inductance of the coil will change as a tooth on the gear positions itself directly below the coil. This change in inductance will be monitored and converted into a pulse and counted for some interval of time. If one knows the number of teeth on the gear and the number of pulses occurring in some interval of time, the angular velocity can be found.

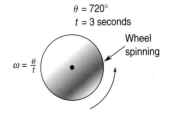

Figure 6.12
A coil used as a proximity sensor functioning as a tachometer.

 ANGULAR VELOCITY CALCULATIONS

Angular velocity is a rate at which an angle changes and is given by:

$$\omega = \frac{\theta}{t} \ (\theta \text{ can be in degrees, revolutions, or radians})$$

Angular velocity = angle/time

EXAMPLE 6.8

A wheel turns 720° in 3 seconds. What is its angular velocity?

$$\theta = 720°$$

$$t = 3 \text{ s}$$

$$\omega = ?$$

$$\omega = \frac{\theta}{t}$$

$$\omega = \frac{720°}{3 \text{ s}}$$

$$\omega = 240 \frac{\text{deg}}{\text{s}}$$

$\theta = 720°$
$t = 3$ seconds

Wheel spinning

$\omega = \frac{\theta}{t}$

Figure 6.13

EXAMPLE 6.9

A top revolves 4.2 radians in 2 seconds. What is its angular velocity?

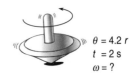

$\theta = 4.2 \ r$
$t = 2 \text{ s}$
$\omega = ?$

Figure 6.14

$$\theta = 4.2 \text{ rad}$$

$$t = 2 \text{ s}$$

$$\omega = \text{?}$$

$$\omega = \frac{\theta}{t}$$

$$\omega = \frac{4.2 \text{ rad}}{2 \text{ s}}$$

$$\omega = 2.1 \frac{\text{rad}}{\text{s}}$$

EXAMPLE 6.10

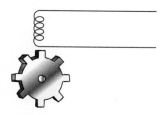

Figure 6.15

Suppose a gear is rotating and being monitored by the tachometer described above. If the gear has ten teeth and produces a count of 500 counts in 1 minute, what is the angular velocity of the gear?

Since the gear has ten teeth, in one complete revolution there will be ten pulses out of the tachometer. Therefore, we want to convert 500 counts into revolutions:

$$500 \text{ counts} \left(\frac{1 \text{rev}}{10 \text{ counts}} \right) = 50 \text{ rev}$$

$$\omega = \frac{\theta}{t}$$

$$\omega = \frac{50 \text{ rev}}{1 \text{ min}} = 50 \text{ rpm}$$

CIRCULAR VELOCITY

We can now calculate the velocity of an object moving in a circular path. Recall that velocity $v = \frac{d}{t}$.

For a circle

$$d = r \cdot \theta$$

$$\omega = \frac{\theta}{t}$$

Therefore:

$$v = \frac{d}{t}$$

$$v = \frac{r \cdot \theta}{t}$$

$$v = r \cdot \omega$$

$$v = r \cdot \omega \quad \leftarrow \quad \text{Velocity of an object moving in a circle}$$
(ω in radians/time) or the velocity of a wheel rolling along a surface (Fig. 6.16).

Velocity = radius × angular velocity

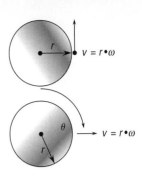

EXAMPLE 6.11

A wheel with a radius of 1.25 ft is rotating at 4.2 rad/s. How fast is the car moving? (See Figure 6.17).

$$r = 1.25 \text{ ft}$$

$$\theta = 4.2\frac{\text{rad}}{\text{s}}$$

$$v = ?$$

$$v = r\omega$$

$$v = (1.25 \text{ ft})\left(4.2\frac{\text{rad}}{\text{s}}\right)$$

$$v = 5.3\frac{\text{ft}}{\text{s}} \quad \text{Note that radians does not appear in the answer because it is dimensionless.}$$

Figure 6.16
Velocity of an object moving in a circle (ω in radians/time) or the velocity of a wheel rolling along a surface can be determined by the equation $v = r \cdot \omega$.

Figure 6.17
The speed of the car depends on how fast the wheels are spinning and their radius.

 ## TRANSMISSIONS

Gears and wheels can be coupled in such a way as to change their speed, direction of rotation, and torque capability (Fig. 6.18).

To determine the speed of rotation of a wheel or gear being driven by a driving wheel or gear, we use the following formulas:

$$N \cdot S = n \cdot s$$
N = angle or angular speed of the driver
S = size of driving diameter or number of teeth
n = angle or angular speed of the driven
s = size of driven diameter or number of teeth

Figure 6.18
Couple gears and wheels together to change speed, direction of rotation, and torque capability.

EXAMPLE 6.12

A wheel of diameter 3 in spinning at 32 rpm is driving a wheel of diameter 6 in. How fast is the driven wheel spinning?

$$S = 3 \text{ in}$$

$$s = 6 \text{ in}$$

$$N = 32 \text{ rpm}$$

$$n = ?$$

$$NS = ns$$

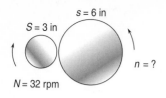

Figure 6.19

$$\frac{NS}{s} = \frac{n\cancel{s}}{\cancel{s}}$$

$$n = \frac{NS}{s}$$

$$n = \frac{(32 \text{ rpm})(3 \cancel{\text{ in}})}{6 \cancel{\text{ in}}}$$

$$n = 16 \text{ rpm}$$

EXAMPLE 6.13

If a wheel that must spin at 3,600 rpm is being driven by a wheel of diameter 2 cm spinning at 180 rpm, what diameter should the driven wheel have?

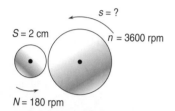

Figure 6.20

$$S = 2 \text{ cm}$$
$$N = 180 \text{ rpm}$$
$$n = 3,600 \text{ rpm}$$
$$s = ?$$
$$NS = ns$$
$$\frac{NS}{n} = \frac{\cancel{n}s}{\cancel{n}}$$
$$s = \frac{NS}{n}$$
$$s = \frac{(180 \cancel{\text{ rpm}})(2 \text{ cm})}{3600 \cancel{\text{ rpm}}}$$
$$s = 0.1 \text{ cm}$$

A distance can separate the driving and driven wheels or gears by using a belt or chain, and the two wheels or gears now rotate in the same direction (Fig. 6.21).

Two wheels can be made to rotate in the opposite direction if the belt is twisted (Fig. 6.22).

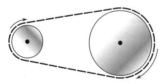

Figure 6.21
A distance can separate the driving and driven wheels or gears by using a belt or chain, and the two wheels or gears now rotate in the same direction.

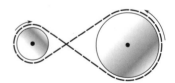

Figure 6.22
Two wheels can be made to rotate in the opposite direction if the belt is twisted.

EXAMPLE 6.14

A gear with 80 teeth spinning at 100 rpm is driving a gear with 20 teeth. How fast is the driven gear spinning?

$$S = 80 \text{ teeth}$$

$$s = 20 \text{ teeth}$$

$$N = 100 \text{ rpm}$$

$$n = ?$$

$$NS = ns$$

$$\frac{NS}{s} = \frac{n\cancel{s}}{\cancel{s}} \quad \leftarrow \quad \text{solving for n}$$

$$n = \frac{NS}{s}$$

$$n = \frac{(100 \text{ rpm})(80 \text{ teeth})}{20 \text{ teeth}}$$

$$n = 400 \text{ rpm}$$

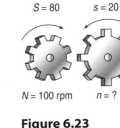

S = 80   s = 20

N = 100 rpm   n = ?

Figure 6.23

Multiple wheels and gears can be connected to create multiple speeds and rotations (Fig. 6.24). To solve problems such as these, apply these equations to two wheels or gears at a time.

Figure 6.24
Multiple wheels and gears can be connected to create multiple speeds and rotations.

 ## TORQUE

Everyone who has ever used a wrench knows that it is easier to turn a wrench with a long handle than a short one. The longer wrench has more torque (Fig. 6.25).

Torque is defined mathematically as:

$$T = F \cdot r$$
$$\text{Torque} = \text{force} \times \text{radius arm}$$

Torque has units of foot-pounds or Newton meters. The force is applied perpendicular to the radius arm as shown in Fig. 6.26.

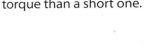

Figure 6.25
A longer wrench has more torque than a short one.

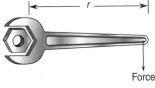

Figure 6.26
Force is applied perpendicular to the radius arm to generate the torque.

EXAMPLE 6.15

A torque wrench measures the torque applied to a bolt (see Fig. 6.27). Suppose a torque wrench handle is 1 ft long and that 100 Lb is applied to the handle. What does the torque wrench read?

$$F = 100 \text{ Lb}$$

$$r = 1 \text{ ft}$$

$$T = ?$$

Figure 6.27
A mechanic's torque wrench.

$$T = F \cdot r$$
$$T = (100 \text{ Lb})(1 \text{ ft})$$
$$T = 100 \text{ ft Lb}$$

Torque and Power

Every motor has horsepower and torque as two of its specifications. *Torque* is a measure of a motor's strength to rotate something connected to its shaft some distance from the center of rotation. Torque and power are related by:

$$P = T \cdot \omega$$
Power = torque × angular velocity (in radians/time)

EXAMPLE 6.16

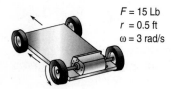

$F = 15$ Lb
$r = 0.5$ ft
$\omega = 3$ rad/s

Figure 6.28

A motor connected to a wheel drives a cart. The cart requires 15 Lb of force to move it. The wheel has a radius of 0.5 ft and is rotating at 3.0 rad/s. a) What torque is the motor putting out? b) What power is the motor generating? c) What is this power in horsepower and watts?

a) $F = 15$ Lb

$r = 0.5$ ft

$\omega = 3.0 \dfrac{\text{rad}}{\text{s}}$

$1 \text{ hp} = 550 \dfrac{\text{ft lb}}{\text{s}}$

$1 \text{ hp} = 746 \text{ W}$

$T = ?$

$T = F \cdot r$

$T = (15 \text{ Lb})(0.5 \text{ ft}) = 7.5 \text{ ft Lb}$

b) $P = ?$

$P = T \cdot \omega$

$P = (7.5 \text{ ft Lb})\left(3 \dfrac{\text{rad}}{\text{s}}\right) = 23 \dfrac{\text{ft Lb}}{\text{s}}$

c) Converting foot-pounds into horsepower

$23 \dfrac{\text{ft Lb}}{\text{s}}\left(\dfrac{1\text{hp}}{550 \dfrac{\text{ft Lb}}{\text{s}}}\right) = 0.042 \text{ hp}$

Converting horsepower into watts

$0.042 \text{ hp}\left(\dfrac{746 \text{ W}}{1\text{hp}}\right) = 31 \text{ W}$

Changing Torque

Sometimes it is necessary to increase the torque delivered to a wheel, for example, when driving up a hill. The torque can be changed in multiple wheel or gear systems by changing the relative size of the driving and driven gears or wheels (Fig. 6.29) according to:

$$\frac{T_d}{T_D} = \frac{s}{S}$$

T_D = driving torque
T_d = driven torque
S = size of driving diameter or number of teeth
s = size of driven diameter or number of teeth

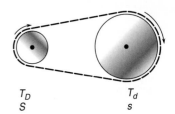

Figure 6.29
The torque can be changed in multiple wheel or gear systems by changing the relative size of the driving and driven gears or wheels.

EXAMPLE 6.17

What is the increase in torque when the driving torque is 12 ft Lb, the driving wheel diameter is 0.16 ft, and the driven wheel is 2 ft in diameter?

$$T_D = 12 \text{ ft Lb}$$

$$S = 0.16 \text{ ft}$$

$$s = 2.0 \text{ ft}$$

$$T_d = ?$$

$$\frac{T_d}{T_D} = \frac{s}{S}$$

$$T_d = T_D \frac{s}{S}$$

$$T_d = 12 \text{ ft Lb} \left(\frac{2.0 \text{ ft}}{0.16 \text{ ft}} \right)$$

$$T_d = 150 \text{ ft Lb}$$

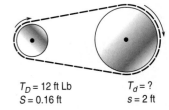

$T_D = 12 \text{ ft Lb}$ $T_d = ?$
$S = 0.16 \text{ ft}$ $s = 2 \text{ ft}$

Figure 6.30

CENTRIPETAL FORCE

When driving a car around a corner, you feel yourself being forced to the passenger side. The faster or the tighter you make the turn, the greater the force. This force is called *centripetal force*. The origin of this force can be explained as follows. When driving a car straight, your body has momentum straight ahead. When making a turn, in order for your body to move with the car, you must hold on to the wheel tightly to "force" your body into this new direction, otherwise you would move in a straight path. This force is the centripetal force (Fig. 6.31).

Remember, changing the motion of something means acceleration, and $F = m \times a$. If a body is being accelerated into a circular path, it can be shown that the

Figure 6.31
Centripetal force generated by changing the motion of a body into a circular path.

acceleration toward the center of the circle is $a = r\omega^2$. Therefore the centripetal force is given:

$$F_c = m \cdot r \cdot \omega^2$$

Force_centripetal = mass × radius × (angular velocity(rad/s))²

EXAMPLE 6.18

What is the centripetal force on a shoe in a washing machine if the shoe has a mass of 0.35 kg and the inside of the machine is rotating at 120 rpm and has a radius of 0.30 m?

$m = 0.35$ kg
$r = 0.3$ m
$\omega = 120$ rpm
$F_c = ?$

Figure 6.32

$$m = 0.35 \text{ kg}$$

$$\omega = 120 \frac{\text{rev}}{\text{min}}$$

Converting ω to rads per second

$$120 \frac{\text{rev}}{\text{min}}\left(\frac{2\pi \text{ rad}}{1 \text{ rev}}\right)\left(\frac{1 \text{ min}}{60 \text{ s}}\right) = 13 \frac{\text{rad}}{\text{s}}$$

$$F_c = ?$$

$$F_c = m \cdot r \cdot \omega^2$$

$$F_c = 0.35 \text{ kg} \cdot 0.30 \text{ m} \cdot \left(13 \frac{\text{rad}}{\text{s}}\right)^2$$

$$F_c = 18 \text{ N} \quad \leftarrow \quad \text{This is about 4 Lb of force.}$$

CHAPTER SUMMARY

1 rev = 360° = 2π rad

$d = r \cdot \theta \quad \leftarrow \quad \theta$ must be in radians

Distance = radius × angle (in radians)

If a wheel of radius r rolls along a surface through an angle θ, the wheel has traveled a distance d, where $d = r \cdot \theta \quad \leftarrow \quad \theta$ must be in radians

$\omega = \dfrac{\theta}{t}$ (θ can be in degrees, revolutions, or radians)

Angular velocity = angle/time

$$v = \frac{d}{t} = \frac{r \cdot \theta}{t} = r \cdot \omega$$

$v = r \cdot \omega \quad \leftarrow \quad$ Velocity of an object moving in a circle (ω in radians/time) or the velocity of a wheel rolling along a surface

Velocity = radius × angular velocity

$N \cdot S = n \cdot s$

N = angle or angular speed of the driver

S = size of driving diameter or number of teeth

n = angle or angular speed of the driven

s = size of driven diameter or number of teeth

$$T = F \cdot r$$

Torque = force × radius arm

Torque has units of foot-pounds or Newton meters

$$P = T \cdot \omega$$

Power = torque × angular velocity
(in radians/time)

$$\frac{T_d}{T_D} = \frac{s}{S}$$

T_D = driving torque

T_d = driven torque

S = size of driving diameter or number of teeth

s = size of driven diameter or number of teeth

$$F_c = m \cdot r \cdot \omega^2$$

$$\text{Force}_{\text{centripetal}} = \text{mass} \times \text{radius} \times (\text{angular velocity(rad/s)})^2$$

Problem-Solving Tips

- Always use angular units of radians when calculating distance (d), velocity (v), power (P), and force$_{\text{centripetal}}$ (F_c).
- In transmission problems, capital letters refer to the drivers, lowercase to the driven.

PROBLEMS

1. A point on a CD rotates through an angle of 1 rev. The radius of the CD is 2.25 in. How far has this point traveled? Remember that θ must be in radians.

2. Suppose you are fixing a broken CD player because it is not spinning fast enough. You measure that the CD completes 5 revolutions in 1 second. What is the angular velocity in revolutions per minute?

3. You are asked to design a robot that moves at a speed of 120 in/min. The drive wheel is connected to a motor spinning at 25 rpm. What should the radius of the wheel be?

4. A wheel of diameter 3 in is being driven by another wheel of diameter 6 in. How many times faster is the driven wheel spinning compared with the driving wheel?

5. A gear with 100 teeth is being driven by another gear. The driven gear is spinning at 10 rpm, and the driving gear is spinning at 5 rpm. How many teeth does the driven gear have?

6. A wrench with an 8-inch handle is used to tighten a bolt. Twenty-two pounds of force is applied to the handle. What is the torque in ft Lb applied to the bolt?

7. A motor drives a conveyer belt. If the motor's angular speed is 1,870 rpm and its torque is 35 ft Lb, what is its horsepower? Remember that 1 hp = 550 ft lb/s.

8. Using two gears, how can you double the torque of a motor?

9. A motor of torque 12 ft Lb is driving a wheel of 2 in in diameter. This wheel is connected by a belt to another driven wheel of 1 in diameter. What is the torque of the 1-in wheel?

10. What is the angular velocity of the earth on its own axis?

11. A floppy disk in a computer rotates at 31.4 rad/s. How many revolutions does it make in a 30-second reading of the disk? (Be careful of the units!)

12. Suppose you wanted to build an elevator to outer space by attaching a 5.98×10^{24}-kg mass in space to a cable 22,000 miles long and drop the other end of the cable to the earth and anchor it to the ground. The centripetal force would keep the cable taut, and you could ride up the cable to space. What would the centripetal force be on the cable? Why doesn't NASA do this?

13. Show that $m \cdot r \cdot \omega^2 = m\dfrac{v^2}{r}$

14. To prevent a car from skidding on a turn, the frictional force of the tire on the road must be greater than or equal to the centripetal force. How fast can the car make a turn if the coefficient of the tire with the road is $\mu = 0.9$, the car weighs 9345 N, and the radius of the turn is 25 m. Hint: Maximum speed occurs when $F_c = F_f$.

15. If you build an electric cart and the drive motor that rotates at 1,750 rpm is connected to a drive wheel of radius 6.0 in, how fast will this cart move?

Machines

Figure 7.1

This chapter describes the basic types of machines. *Machines* are devices that help us perform work that could not otherwise be performed. Typically, they generate more force than we are capable of generating and are found in many types of electromechanical equipment.

Theoretically, machines work on the principle:

Work in = work out

The amount of work you put into the machine equals the amount you get out.

Note that work in = work out is only an idealization because there will always be friction that just wastes energy.

Rewriting this equation using the equation for work, $W = F \cdot d$:

$$F_{in} \cdot d_{in} = F_{out} \cdot d_{out}$$

Force in × distance in = Force out × distance out

This equation says that when you apply a force to a part of the machine and move it some distance (d), another part of the machine will generate a force out and move it a distance out. All machines work on this principle. Figure 7.2 illustrates the six basic machines.

Most machines you are familiar with are *compound machines,* that is, they consist of one or more of these basic machines (Fig. 7.3).

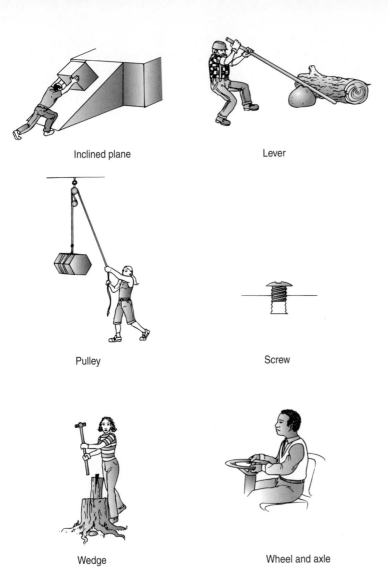

Inclined plane

Lever

Pulley

Screw

Wedge

Wheel and axle

Figure 7.2
The six basic machines.

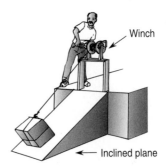

Winch

Inclined plane

Figure 7.3
A compound machine consists of
several basic machines.

 MECHANICAL ADVANTAGE

Mechanical advantage (MA) is a measure of how many times your strength is multiplied by using a machine. Because a machine is not 100% efficient, some of the work going in is wasted because of friction. How much force a machine can actually generate is defined by:

$$\text{Mechanical advantage} \quad \rightarrow \quad MA = \frac{F_o}{F_i} = \frac{\text{force out}}{\text{force in}}$$

If the force generated by the machine is large compared with the input force, the MA is large.

EXAMPLE 7.1

Using a machine to generate 2,000 Lb with 20 Lb of force, what is the MA?

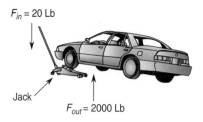

F_{in} = 20 Lb

Jack

F_{out} = 2000 Lb

Figure 7.4

$$F_{out} = 2000 \text{ Lb}$$

$$F_{in} = 20 \text{ Lb}$$

$$MA = ?$$

$$MA = \frac{F_o}{F_i}$$

$$MA = \frac{2,000 \text{ Lb}}{20 \text{ Lb}} = 100 \leftarrow \text{This machine can increase your strength 100 times.}$$

EXAMPLE 7.2

A machine has an MA of 20. What weight can be moved with an applied force of 5 Lb?

$$MA = 20$$

$$F_i = 5.0 \text{ Lb}$$

$$F_o = ?$$

$$MA = \frac{F_o}{F_i}$$

$$F_o = MA \cdot F_i$$

$$F_o = 20 \cdot 5 \text{ Lb}$$

$$F_o = 100 \text{ Lb}$$

 ## EFFICIENCY

The *efficiency* of a machine is a measure of how much work it puts out versus how much work goes in. It will always be less than 100% because of friction.

$$\text{Efficiency} = \frac{\text{work out}}{\text{work in}}$$

EXAMPLE 7.3

A machine puts out 89 J of work for an input of 98 J. What is its efficiency?

Work in = 98 J

Work out = 89 J

$$\text{Efficiency} = \frac{\text{work out}}{\text{work in}}$$

$$\text{Efficiency} = \frac{89 \text{ J}}{98 \text{ J}}$$

$$\text{Efficiency} = 0.91 \quad \leftarrow \quad \text{The machine is 91\% efficient.}$$

Applying input force to a machine using a crank or a handle and moving that implement a distance larger than that of the output of the machine will give you a very large MA. Think about jacking up a car. You have to pump the jack handle through a large distance to make the car move just a little (Fig. 7.5), but with that little force you are able to lift a car!

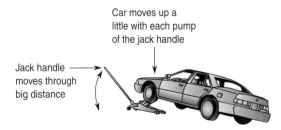

Car moves up a little with each pump of the jack handle

Jack handle moves through big distance

Figure 7.5
When jacking a car, you must pump the jack handle a big distance to move the car up just a little.

From the definition of efficiency, the MA can also be expressed another way:

$$\text{MA} = \text{efficiency} \times \frac{d_i}{d_o}$$

$$\text{Mechanical advantage} = \text{efficiency} \times \frac{\text{effort distance in}}{\text{load distance out}}$$

EXAMPLE 7.4

When jacking up a car, suppose the jack handle moves through a distance of 36 in and the car moves up 0.5 in. What is the mechanical advantage? Assume a 40% efficiency.

$$d_i = 36 \text{ in}$$

$$d_o = 0.50 \text{ in}$$

$$\text{Efficiency} = 0.4$$

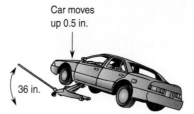

Car moves
up 0.5 in.

36 in.

Figure 7.6

$$MA = ?$$

$$MA = \text{efficiency} \times \frac{d_i}{d_o}$$

$$MA = 0.40 \times \frac{36 \text{ in}}{0.50 \text{ in}}$$

$$MA = 29 \quad \leftarrow \quad \text{Your strength has increased 29 times!}$$

THE LEVER

There are three types of levers: class 1, class 2, and class 3 (Figs. 7.7–7.9).

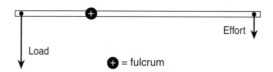

Load

$\oplus$ = fulcrum

Effort

Figure 7.7
A class 1 lever.

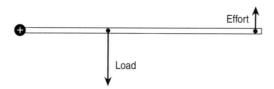

Effort

Load

Figure 7.8
A class 2 lever.

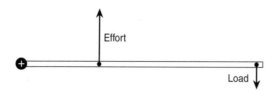

Effort

Load

Figure 7.9
A class 3 lever.

The lever has little friction. Therefore the efficiency is 100% and MA $= \dfrac{d_i}{d_o}$.

EXAMPLE 7.5

A lever of class 1 is used to lift a 100 Lb weight placed 2 ft from the fulcrum. The input force is placed 4 ft from the fulcrum. What is the MA of this lever, and how much input force is needed to raise the load?

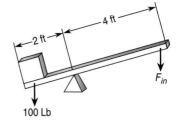

Figure 7.10

$$d_i = 4 \text{ ft}$$

$$d_o = 2 \text{ ft}$$

$$F_{out} = 100 \text{ Lb}$$

$$F_{in} = ?$$

$$\text{MA} = ?$$

$$\text{MA} = \frac{d_i}{d_o}$$

$$\text{MA} = \frac{4 \text{ ft}}{2 \text{ ft}} = 2$$

$$\text{MA} = \frac{F_o}{F_i}$$

$$\text{MA} \cdot F_i = \left(\frac{F_o}{\cancel{F_i}}\right)\cancel{F_i} \quad \leftarrow \quad \text{solving for } F_{in}$$

$$F_o = \text{MA} \cdot F_i$$

$$\frac{F_o}{\text{MA}} = \frac{\cancel{\text{MA}} \cdot F_i}{\cancel{\text{MA}}}$$

$$F_i = \frac{F_o}{\text{MA}}$$

$$F_i = \frac{100 \text{ Lb}}{2} = 50 \text{ Lb}$$

Electromechanical devices use levers. See Fig. 7.11.

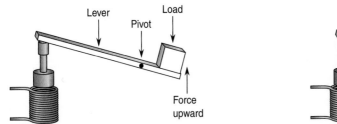
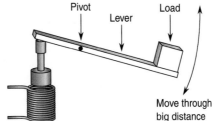

Figure 7.11
A class 1 lever is sometimes used with a solenoid either to generate more force or to move something through a greater distance.

EXAMPLE 7.6

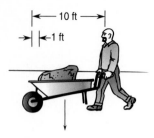

← 10 ft →

→| |← 1 ft

Figure 7.12

A wheelbarrow is a class 2 lever and is used to lift a 500-Lb weight placed 1 ft from the fulcrum. The input force is placed 10 ft from the fulcrum. What is the MA of this lever, and how much input force is needed to raise the load?

$$d_i = 10 \text{ ft}$$

$$d_o = 1 \text{ ft}$$

$$F_o = 500 \text{ Lb}$$

$$F_i = ?$$

$$MA = ?$$

$$MA = \frac{d_i}{d_o}$$

$$MA = \frac{10 \text{ ft}}{1 \text{ ft}} = 10$$

$$MA = \frac{F_o}{F_i}$$

$$MA \cdot F_i = \left(\frac{F_o}{F_i}\right) F_i \qquad \leftarrow \quad \text{solving for } F_i$$

$$\frac{MA \cdot F_i}{MA} = \frac{F_o}{MA}$$

$$F_i = \frac{F_o}{MA}$$

$$F_i = \frac{500 \text{ Lb}}{10} = 50 \text{ Lb}$$

EXAMPLE 7.7

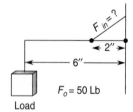

$F_{in} = ?$

← 2″ →

← 6″ →

$F_o = 50 \text{ Lb}$

Load

Figure 7.13

A class 3 lever is used to lift a 50.0-Lb weight placed 6.0 in from the fulcrum. The input force is placed 2.0 in from the fulcrum. What is the MA of this lever, and how much input force is needed to raise the load?

$$d_i = 2 \text{ in}$$

$$d_o = 6 \text{ in}$$

$$F_o = 50 \text{ Lb}$$

$$F_i = ?$$

$$MA = ?$$

$$MA = \frac{d_i}{d_o}$$

$$MA = \frac{2 \text{ in}}{6 \text{ in}} = \frac{1}{3}$$

$$MA = \frac{F_o}{F_i}$$

$$MA \cdot F_i = \left(\frac{F_o}{\cancel{F_i}}\right)\cancel{F_i} \quad \leftarrow \quad \text{solving for } F_i$$

$$F_o = MA \cdot F_i$$

$$\frac{F_o}{MA} = \frac{\cancel{MA} \cdot F_i}{\cancel{MA}}$$

$$F_i = \frac{F_o}{MA}$$

$$F_i = \frac{50 \text{ Lb}}{1/3} = 150 \text{ Lb}$$

THE PULLEY

Ropes suspending weight will have the following tensions in them (see Figure 7.14).

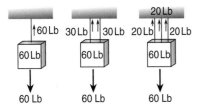

Figure 7.14

If we make the ropes movable by using pulleys, we can lift the load and still have the reduced tension in the strands of rope. The bearings will have some friction in them, therefore:

> If the input force moves in the direction of the load, then:
>
> MA = bearing efficiency × number of strands
>
> If the input force moves in the opposite direction of the load, then:
>
> MA = bearing efficiency × (number of strands − 1)

See Fig. 7.15.

These machines are sometimes called a *block and tackle.* To build one, wind the rope from outside in, always winding either clockwise or counterclockwise (Fig. 7.16).

E X A M P L E   7 . 8

A 300-Lb engine is being lifted out of a car with a block winch, shown in Figure 7.17 on the next page. How hard does the mechanic have to pull to lift the engine? Assume an efficiency of 90%.

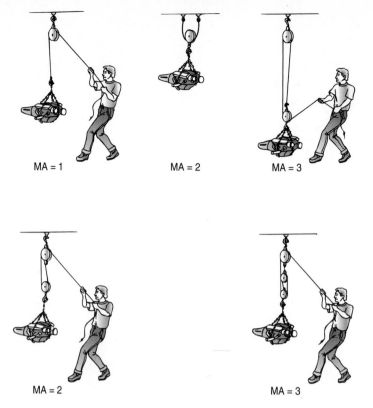

MA = 1   MA = 2   MA = 3

MA = 2   MA = 3

Figure 7.15

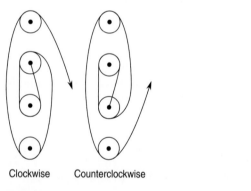

Clockwise   Counterclockwise

Figure 7.16
To a wind a block and
tackle, wind the rope from
outside in, always winding
either clockwise or
counterclockwise.

Figure 7.17

There are four strands, with the last one being pulled in the opposite direction of the engine, therefore:

$$MA = \text{efficiency} \times (\text{number of strands} - 1)$$

$$MA = 0.90 \times (4 - 1)$$

$$MA = 0.90 \times 3$$

$$\text{MA} = 2.7$$

$$\text{MA} = \frac{F_o}{F_i}$$

$$F_i = \frac{F_o}{\text{MA}}$$

$$F_i = \frac{300 \text{ Lb}}{2.7}$$

$$F_i = 111 \text{ Lb}$$

THE WHEEL AND AXLE

A *wheel and axle* machine converts a large rotary motion into a small one, but it increases the output force (Fig. 7.18).

Screwdriver   Winch

Bicycle sprocket   Steering wheel

TV knob

Figure 7.18
A steering wheel, a bicycle sprocket and pedals, a screwdriver, a winch, and a knob on a TV are examples of a wheel and axle machine.

The distance in (d_i) and the distance out (d_o) is the radius or diameter of the wheel and axle.

$$MA = \text{bearing efficiency} \times \frac{d_i}{d_o}$$

$$MA = \text{bearing efficiency} \times \frac{\text{wheel diameter or radius}}{\text{axle diameter or radius}}$$

EXAMPLE 7.9

A winch handle is 12 in long and rotates a drum, which is 2 in in radius. The load is 60 Lb. The efficiency of the bearings is 95%. What is the mechanical advantage, and how much input force is needed to lift the weight?

$$d_i = 12 \text{ in}$$

$$d_o = 2 \text{ in}$$

$$F_o = 60 \text{ Lb}$$

$$\text{Efficiency} = 92\%$$

$$F_i = ?$$

$$MA = ?$$

$$MA = \text{efficiency} \times \frac{d_i}{d_o}$$

$$MA = 0.92 \times \frac{12 \, \cancel{\text{in}}}{2 \, \cancel{\text{in}}} = 5.5$$

$$MA = \frac{F_o}{F_i}$$

$$MA \cdot F_i = \left(\frac{F_o}{\cancel{F_i}}\right)\cancel{F_i} \quad \leftarrow \quad \text{solving for } F_i$$

$$F_o = MA \cdot F_i$$

$$\frac{F_o}{MA} = \frac{\cancel{MA} \cdot F_i}{\cancel{MA}}$$

$$F_i = \frac{F_o}{MA}$$

$$F_i = \frac{60 \text{ Lb}}{5.5} = 11 \text{ Lb}$$

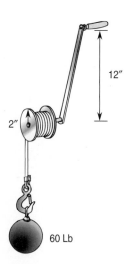

Figure 7.19

12″

2″

60 Lb

▪ THE INCLINED PLANE

Sliding a heavy object up an inclined plane takes less force than actually picking it up and raising it to that height. See Fig. 7.20.

An example of an inclined plane is a cam shaft in an engine. See Fig. 7.21.

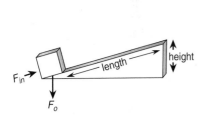

Figure 7.20

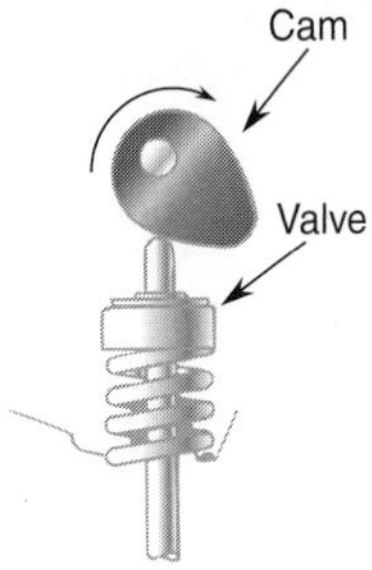

Cam

Valve

Figure 7.21
A cam is an example of an inclined plane.

For an inclined plane, MA is:

$$MA = efficiency \times \frac{length}{height}$$

EXAMPLE 7.10

A 75-lb crate is slid up a greased inclined plane that is 12.0 ft long and 3.0 ft high. What is the MA, and how much input force is needed? Assume an efficiency of 65%.

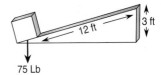

Figure 7.22

$$Length = 12.0 \text{ ft}$$

$$Height = 3.0 \text{ ft}$$

$$F_o = 75 \text{ Lb}$$

$$Efficiency = 0.65$$

$$F_i = ?$$

$$MA = ?$$

$$MA = Efficiency \times \frac{length}{height}$$

$$MA = 0.65 \times \frac{12.0 \text{ ft}}{3.0 \text{ ft}}$$

$$MA = 2.6$$

$$MA = \frac{F_o}{F_i}$$

$$MA \cdot F_{in} = \left(\frac{F_o}{\cancel{F_i}}\right)\cancel{F_i} \quad \leftarrow \quad \text{solving for } F_i$$

$$F_o = MA \cdot F_i$$

$$\frac{F_o}{MA} = \frac{\cancel{MA} \cdot F_i}{\cancel{MA}}$$

$$F_i = \frac{F_o}{MA}$$

$$F_i = \frac{75 \text{ Lb}}{2.6} = 29 \text{ Lb}$$

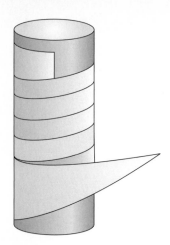

Figure 7.23
A screw is an inclined plane wrapped around a cylinder.

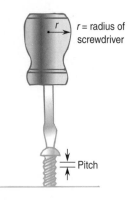

r = radius of screwdriver

Pitch

Figure 7.24
The pitch of a screw is the distance between adjacent threads.

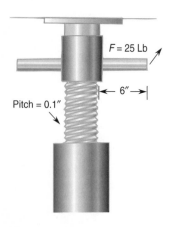

F = 25 Lb

6″

Pitch = 0.1″

Figure 7.25

THE SCREW

A *screw* is essentially an inclined plane wrapped around a cylinder (Fig. 7.23).

The pitch of a screw is the distance between adjacent threads (Fig. 7.24).

When the screw is rotated through one complete revolution, it rises up one pitch distance. A handle on a screwdriver or a handle mounted on the screw allows the screw to be turned with ease. Lubrication will maximize efficiency. The mechanical advantage is:

$$MA = \text{efficiency} \times \frac{2\pi r}{\text{pitch}}$$

$$MA = \text{efficiency} \times \frac{2\pi \cdot \text{handle radius}}{\text{pitch}}$$

EXAMPLE 7.11

A jackscrew has a pitch of 0.10 in and a handle that is 6.0 in in length (Fig. 7.25). What is the MA, and how much weight can it lift with 25 Lb of input force? Assume 85% efficiency.

$$r = 6.0 \text{ in}$$
$$\text{pitch} = 0.10 \text{ in}$$
$$F_i = 25 \text{ Lb}$$
$$\text{Efficiency} = 0.85$$
$$F_o = ?$$
$$MA = ?$$

$$MA = \text{efficiency} \times \frac{2\pi r}{\text{pitch}}$$

$$MA = 0.85 \times \frac{2\pi(6.0 \text{ in})}{0.10 \text{ in}}$$

$$MA = 320$$

$$MA = \frac{F_o}{F_i}$$

$$MA \cdot F_i = \left(\frac{F_o}{F_i}\right) F_i \quad \leftarrow \quad \text{solving for } F_o$$

$$F_o = MA \cdot F_i$$

$$F_o = 320 \cdot 25 \text{ Lb}$$

$$F_o = 8,\overline{0}00 \text{ Lb}$$

Screws have an enormous mechanical advantage!

WEDGE

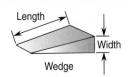
Length
Width
Wedge

Wedges are generally used as an aid in splitting things apart.

A wedge typically has a lot of friction, and therefore its efficiency is very low (Fig. 7.26). We can approximate the MA of the wedge by:

$$MA = \text{efficiency} \times \frac{\text{length}}{\text{width}}$$

Axe

Shears

EXAMPLE 7.12

What is the mechanical advantage of a wedge with a length of 9 in, a width of 1 in, and an efficiency of 30%? See Fig. 7.27.

$$\text{Length} = 9 \text{ in}$$
$$\text{Width} = 1 \text{ in}$$
$$\text{Efficiency} = 30\%$$
$$MA = 0.30 \times \frac{9 \text{ in}}{1 \text{ in}}$$
$$MA = 3$$

Figure 7.26
A wedge typically has a lot of friction and very low efficiency.

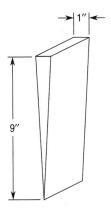
1″
9″

COMPOUND MACHINES

Figure 7.27

Using two or more basic machines at the same time results in a much higher *MA*. The *MA* of a compound machine is:

$$MA_{Total} = MA_1 \times MA_2 \times MA_3 \times \ldots$$

EXAMPLE 7.13

A compound machine consists of a 6-strand pulley system with an efficiency of 88%, a wheel and axle with the wheel radius of 4.0 in and the axle radius of 2.0 in, and an efficiency of 98%. What is the MA of this compound machine?

For the pulley:

$$6 \text{ strands}$$
$$\text{Efficiency} = 0.88$$
$$MA = \text{efficiency} \times \text{number of strands}$$
$$MA = 0.88 \times 6$$
$$MA = 5.3$$

For the wheel and axle:

$$d_i = 4.0 \text{ in}$$
$$d_o = 2.0 \text{ in}$$
$$\text{Efficiency} = 0.98$$

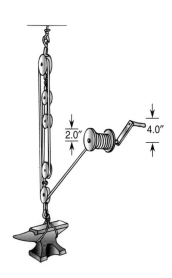
2.0″
4.0″

Figure 7.28

$$MA = \text{efficiency} \times \frac{d_i}{d_o}$$

$$MA = 0.98 \times \frac{4.0 \text{ in}}{2.0}$$

$$MA = 2.0$$

The total MA is:

$$MA_{Total} = MA_1 \times MA_2$$

$$MA_{Total} = 5.3 \times 2.0$$

$$MA_{Total} = 11$$

CHAPTER SUMMARY

The basic law of machines assumes no friction: Work in = work out

Written another way:

$$F_{in} \cdot d_{in} = F_{out} \cdot d_{out}$$

Force in × distance in = force out × distance out

$$\text{Efficiency} = \frac{\text{work out}}{\text{work in}}$$

Mechanical advantage

$$MA = \frac{F_o}{F_i} = \frac{\text{force out}}{\text{force in}}$$

or

$$MA = \text{efficiency} \times \frac{d_i}{d_o}$$

Mechanical advantage

$$= \text{efficiency} \times \frac{\text{effort distance in}}{\text{load distance out}}$$

There are six basic machines: the lever, pulley, wheel and axle, inclined plane, screw, and the wedge.

1) The lever: $MA = \text{efficiency} \times \dfrac{d_i}{d_o}$

2) The pulley:
If the input force moves in the direction of the load, then MA = efficiency × number of strands
If it moves in the opposite direction of the load, then MA = efficiency × (number of strands −1)

3) The wheel and axle: $MA = \text{efficiency} \times \dfrac{d_i}{d_o}$

4) The inclined plane: $MA = \text{efficiency} \times \dfrac{\text{length}}{\text{height}}$

5) The screw: $MA = \text{efficiency} \times \dfrac{2\pi r}{\text{pitch}}$

6) The wedge: $MA = \text{efficiency} \times \dfrac{\text{length}}{\text{width}}$

For compound machines: $MA_{Total} = MA_1 \times MA_2 \times MA_3 \times \ldots$

Problem-Solving Tips

■ If you have trouble determining which formula to use, remember that MA is a measure of how many times your strength is increased by a machine. It is determined by:

$$MA = \text{efficiency} \times$$

$$\frac{\text{the distance you move a handle, lever, etc. through}}{\text{the distance the machine moves what it's working on}}$$

PROBLEMS

1. A machine is able to lift 500 Lb with 5 Lb of applied force. What is the *MA* of this machine?

2. With an applied force of 25 Lb a machine lifts 100 Lb. Suppose the applied force is moved through a distance of 10 in and the machine lifts

a load 2 in. What are the MA, and the efficiency of this machine?

3. A wheelbarrow with a 6-ft handle lifts 300 Lb of weight placed 1 ft from the wheel. What is the MA of this machine, assuming a 98% efficiency? How much force is needed to lift this load?

4. A car stuck in the mud can be lifted up with a lever. The car is 2 ft from the fulcrum, and the applied force is 10 ft from the fulcrum. If a person stands on the lever with 180 Lb of body weight, how hard is the car pushed upward? Assume 100% efficiency.

5. A body builder is working her arms. Her bicep is attached 2.0 in from the elbow joint. She holds a 55 Lb weight in her hands at a distance of 12 in from the elbow. What is the MA of her bicep, and how much force does the bicep have to generate to lift this weight? Assume 100% efficiency.

6. A block and tackle with five strands is used to lift a 200-Lb motor onto a roof. The motor rises upward while the rope is pulled downward. What force is needed to lift this motor? Assume an efficiency of 92%.

7. A bicycle pedal is 10 in from the center of rotation and is connected to a gear 6 in in radius. What is the MA of this machine? Assume an efficiency of 96%.

8. When building the pyramids in Egypt, workers pulled heavy stone blocks up the side of the pyramids by using skids. If the blocks weighed 4500 Lb and the pyramid was sloped at 40°, how hard did the workers have to push? Assume an efficiency of 62%. How could a pulley system be incorporated to increase the MA?

9. How much force can a nutcracker generate when the handle is 10 in long, the thread pitch is 0.25 in, and the applied force is 4 Lb? Assume an efficiency of 84%.

10. A compound machine consists of a lever with an efficiency of 100%, a block and tackle with an efficiency of 94%, and a wheel and axle with an efficiency of 94% (having MAs of 3, 4, and 5, respectively). What is the total MA for this compound machine? If 20 N of force is applied, what is the output force?

11. A knob on an oscilloscope has a radius of 0.50 in and turns a rotary switch it is connected to with a radius of 0.125 in. What is the MA of this knob? Assume an efficiency of 99%.

12. Suppose an elevator has to exert a force of 435 N to move a crate straight up a distance of 3.0 m. If an inclined plane were used, less force would be needed to raise it the same distance, but it would have to be pulled farther along the inclined plane. How long would this plane need to be if it is to be pulled with only 50.0 N of force? Assume an efficiency of 74%.

13. Design a compound machine that could have aided the ancient Egyptians in building a pyramid. The machine should incorporate five of the basic machines and have an MA of 10,000. Show the dimensions of each machine and its respective MA.

14. An axe head has a length of 20.0 cm and a width of 5.0 cm. If a hammer delivers a blow of 150 N to the head, how much force is delivered to the wood? Assume an efficiency of 34%.

15. How could you redesign your bicep muscle so that you could win arm-wrestling contests more easily?

16. Connecting a solenoid to a lever can increase its pulling force. Draw a diagram showing how this might be done.

Strength of Materials

This chapter describes mechanical properties of materials. Some properties of solid materials are strength, elasticity, malleability, and ductility. The ever-growing field of materials engineering is dedicated to studying these properties as well as others.

ELASTIC PROPERTIES

Some materials, when stretched or compressed, return to their original shape after the force is removed. One example is a spring.

What determines the strength of material? When a material is stretched, its atoms are pulled apart. When it is compressed, its atoms are pushed together. The forces between the atoms resist this change in length. When the applied force causing the atoms to change separation distance is greater than the electric forces between atoms, then a permanent deformation will occur.

ELASTIC LIMIT

A material's *elastic limit* is the maximum amount it can be stretched or compressed without becoming permanently deformed (Fig. 8.1).

SPRINGS

Springs operate below the elastic limit. They return to their original shape after being stretched or compressed. The force needed to change the length of a spring is shown in *Hooke's law*.

$$F = k \cdot \Delta x$$
Force = spring constant × length change

This equation says the force needed to stretch or compress a spring is equal to how stiff the spring is and how much its length is changed. The spring constant (*k*) has units of force/length.

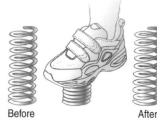
Before　　　　After

Figure 8.1
A material's elastic limit is the maximum amount it can be stretched or compressed without becoming permanently deformed.

EXAMPLE 8.1

A spring has a spring constant of 5 Lb/in and is stretched 2 in. What is the force stretching the spring?

$$k = 5 \, \frac{Lb}{in}$$

$$\Delta x = 2 \text{ in}$$

$$F = k \cdot \Delta x$$

$$F = \left(5 \, \frac{Lb}{\cancel{in}} \right) 2 \cancel{in}$$

$$F = 10 \text{ Lb}$$

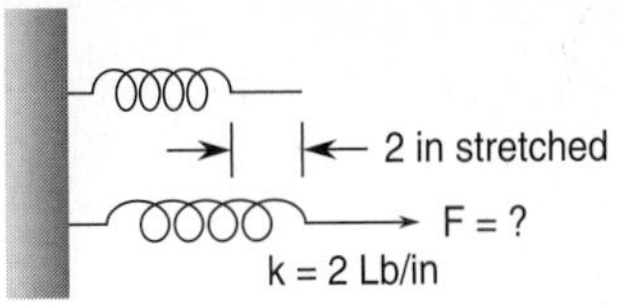

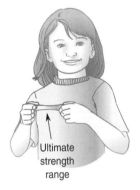

Figure 8.2

2 in stretched
F = ?
k = 2 Lb/in

PLASTIC RANGE

Just beyond the elastic limit is the *plastic range*, where stretching or compressing results in a permanent change in shape (Fig. 8.3). When making objects such as coins or horseshoes, a metal is worked within its plastic range.

Working in the plastic range

Figure 8.3
To permanently form a metal into some shape, you must work in the plastic range of the metal.

Ultimate strength range

Figure 8.4
In the ultimate strength region of a material, stretching the material becomes increasingly easy.

ULTIMATE STRENGTH

At the end of the plastic range, if the material is stretched any farther, the molecules will begin to flow over one another. Stretching the material becomes increasingly easy (Fig. 8.4). The material becomes taffylike in this state. This is called the ultimate strength region.

Table 8.1 lists the elastic limit and ultimate strengths of materials.

Table 8.1

Table of Strengths of Materials

Solid	Elastic limit		Ultimate strength	
	Tension	Compression	Tension	Compression
Aluminum	840	840	1 800	∞
Brass	630	630	2 000	∞
Brick, best hard	30	840	30	840
Brick, common	4	70	4	70
Bronze	2 800	2 800	5 300	∞
Cement, Portland, 1 mo old	28	140	28	140
Cement, Portland, 1 yr old	35	210	35	210
Concrete, Portland, 1 mo old	14	70	14	70
Concrete, Portland, 1 yr old	28	140	28	140
Copper	700	700	2 500	∞
Douglas fir	330	330	500	430
Granite	49	1 300	49	1300
Iron, cast	420	1 800	1 400	5600
Lead	10	10	200	- ∞
Limestone and sandstone	21	630	21	630
Monel metal	6 300	6 300	7 000	∞
Oak, white	310	310	600	520
Pine, white, eastern	270	270	400	345
Slate	35	980	35	980
Steel, bridge cable	6 700	6 700	15 000	∞
Steel, 1 percent C, tempered	5 000	5 000	8 400	8 400
Steel, chrome, tempered	9 100	9 100	11 000	11 000
Steel, stainless	2 100	2 100	5 300	5 300
Steel, structural	2 500	2 500	4 600	4 600

Note: $1 \dfrac{\text{kg}_\text{f}}{\text{cm}^2} = 98.1 \text{ kPa} = 14.22 \dfrac{\text{Lb}}{\text{in}^2}$

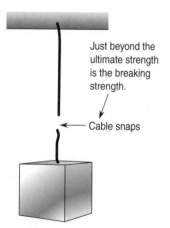

Just beyond the ultimate strength is the breaking strength.

Cable snaps

Figure 8.5
The breaking strength region occurs just before rupture.

BREAKING STRENGTH

Just beyond the ultimate strength range, the material breaks or ruptures (Fig. 8.5).

STRESS

To cause a deformation in a material, stress is applied to the material. *Stress* is the ratio of the applied force to the area it is applied to. Stress has the same units as pressure.

$$\text{Stress} = \frac{F}{A}$$

$$\text{Stress} = \frac{\text{force}}{\text{area}}$$

EXAMPLE 8.2

A car jack is supporting 900 Lb of weight applied to an area of 3 in². What is the stress on the head of the jack?

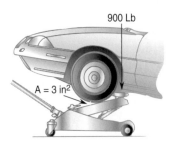

Figure 8.6

$$F = 900 \text{ Lb}$$

$$A = 3 \text{ in}^2$$

$$\text{Stress} = \frac{F}{A}$$

$$\text{Stress} = \frac{900 \text{ Lb}}{3 \text{ in}^2}$$

$$\text{Stress} = 300 \frac{\text{Lb}}{\text{in}^2}$$

There are three types of stress: tensile stress, compressive stress, and shear stress (Figs. 8.7–8.9).

Compressive stress

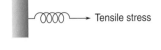 Tensile stress

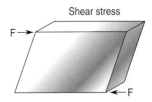

Shear stress

Figure 8.7
Tensile stress is associated with stretching.

Figure 8.8
Compressive stress is associated with squashing or compression.

Figure 8.9
Shear stress is associated with two forces separated from each other, applied in opposite directions. Shear stresses are usually applied to crack something or cut or tear it apart.

Bending consists of both compressive and tensile stress. In Figure 8.10, compression is taking place in the top layers of the materials and a tensile stress in the bottom layers.

 STRAIN

A stress applied to a material results in a strain on the material. Strain is the ratio between the change in length of a material and its original length. Strain has no units.

$$\text{Strain} = \frac{\Delta L}{L}$$

$$\text{Strain} = \frac{\text{length change}}{\text{original length}}$$

Figure 8.10
Bending consists of both compressive and tensile stress.

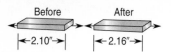

Before   After

|←2.10"→| |←2.16"→|

Figure 8.11

EXAMPLE 8.3

A bar under tension is stretched from 2.10 in to 2.16 in. What is the strain on the bar?

$$\text{Original length} = 2.10 \text{ in}$$

$$\text{Final length} = 2.16 \text{ in}$$

$$\Delta L = 2.16 \text{ in} - 2.10 \text{ in} = 0.060 \text{ in}$$

$$\text{Strain} = \frac{\Delta L}{L}$$

$$\text{Strain} = \frac{0.060 \text{ in}}{2.10 \text{ in}}$$

$$\text{Strain} = 2.9 \times 10^{-4} \quad \leftarrow \quad \text{no unit}$$

There are three types of strain: tensile strain, compressive strain, and shear strain (Figs. 8.12–8.14).

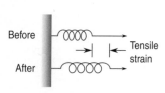

Figure 8.12
Tensile strain is associated with stretching.

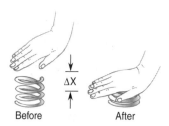

Figure 8.13
Compressive strain is associated with squashing or compression.

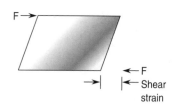

Figure 8.14
Shear strain is associated with two forces separated from each other applied in opposite directions.

The result of stress on a material is strain. If we are working within the elastic limit, the ratio of stress over strain is given by the modulus of elasticity.

Modulus of elasticity: The ratio of the stress over strain.

Young's modulus: Young's modulus applies to tension and compression stresses.

$$Y = \frac{F/A}{\Delta L/L}$$

$$\text{Young's modulus} = \frac{\text{stress}}{\text{strain}}$$

The *modulus of elasticity* is a measure of how much a material changes its shape from an applied force and returns to its original shape after the force is released. Table 8.2 lists the elastic modulus for various materials.

Table 8.2
Elastic Constants for Various Materials in USCS Units

Material	Young's Modulus Y, Lb/in.^{2}	Shear Modulus S, Lb/in.^{2}	Bulk Modulus B, Lb/in.^{2}	Elastic Limit, Lb/in.^{2}	Ultimate Strength, Lb/in.^{2}
Aluminum	10×10^6	3.44×10^6	10×10^6	19,000	21,000
Brass	13×10^6	5.12×10^6	8.5×10^6	55,000	66,000
Copper	17×10^6	6.14×10^6	17×10^6	23,000	49,000
Iron	13×10^6	10×10^6	14×10^6	24,000	47,000
Steel	30×10^6	12×10^6	23×10^6	36,000	71,000

EXAMPLE 8.4

Expansion joints in a bridge allow concrete to thermally expand during the summer. Suppose that a 100-ft section of road expands by 1 in. How much pressure will the road exert on an object stuck between the joint?

$$L = 100.0 \text{ ft}$$

$$\Delta L = 1.0 \text{ in} = 0.083 \text{ ft}$$

$$Y_{\text{concrete}} = 10.6 \times 10^6 \frac{\text{Lb}}{\text{in}^2}$$

$$Y = \frac{F/A}{\Delta L/L}$$

$$F/A = Y \cdot \Delta L/L$$

$$F/A = 10.6 \times 10^6 \frac{\text{Lb}}{\text{in}^2} \cdot \frac{0.083 \text{ ft}}{100.0 \text{ ft}}$$

$$F/A = 8,798 \frac{\text{Lb}}{\text{in}^2}$$

Figure 8.15

A road under this kind of pressure would buckle if an expansion joint was not used.

 ## BULK MODULUS

The *bulk modulus* refers to compression stresses applied to a volume. If a compressive stress is uniformly applied to the volume, the compressive force will cause the volume to shrink.

$$B = -\frac{F/A}{\Delta V/V}$$

$$\text{Bulk modulus} = -\frac{\text{volume stress}}{\text{volume strain}}$$

The minus sign in the equation is there to cancel out the minus sign from the volume change, which is negative, because it shrinks.

EXAMPLE 8.5

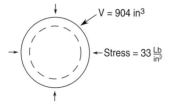

Figure 8.16

Suppose an inflatable ball of original volume 904 in³ is placed under the water with an outside pressure applied to it of 33 Lb/in². If the bulk modulus of the ball is 10^4 Lb/in², what is its change in volume under water?

$$V = 904 \text{ in}^3$$

$$\frac{F}{A} = 33 \text{ Lb/in}^2$$

$$B = 10^4 \text{ Lb/in}^2$$

$$B = -\frac{F/A}{\Delta V/V}$$

$$B = -\frac{\dfrac{F}{A} \cdot V}{\Delta V} \qquad \leftarrow \quad \text{Simplifying the equation}$$

$$\Delta V = -\frac{\dfrac{F}{A} \cdot V}{B} \qquad \leftarrow \quad \text{Solving for } \Delta V \text{ gives:}$$

$$\Delta V = -\frac{33 \dfrac{\cancel{Lb}}{\cancel{in^2}} \cdot 904 \text{ in}^3}{10^4 \dfrac{\cancel{Lb}}{\cancel{in^2}}}$$

$$\Delta V = \frac{33 \cdot 904}{10^4} \text{ in}^3$$

$$\Delta V = 3 \text{ in}^3$$

SHEAR MODULUS

The *shear modulus* applies to a solid under a shearing force. The volume of the object does not change; only its shape changes. The shear modulus is given by:

$$S = \frac{F/A}{d/l}$$

$$\text{Shear modulus} = \frac{\text{shearing stress}}{\text{shearing strain}}$$

HARDNESS

Hardness of a material is determined by the interatomic forces between the atoms resisting a change in separation-distance with an applied force. The hardness of a material can be determined by how easily it scratches. The Brinell method is a test for determining the hardness of a material. It is performed by applying 29,400 N of force to a material with a 10-mm hardened chrome-steel ball. The size of the imprint on the material being tested determines how hard it is (Fig. 8.17).

Some extremely hard materials can be very strong under compression but very weak with a bending or shear force or a tension. For example, brick and tungsten carbide are very strong in the compression mode but not with tensile or shear forces.

 ## MALLEABILITY

The *malleability* of a material is a measure of how easily it can change shape (Fig. 8.18). For example, lead or gold can easily be sculpted because they are very malleable. Malleable materials have a large plastic range when under compression. When these materials are shaped, the atoms tend to slide or roll over one another.

Figure 8.17
A material being tested for hardness.

Table 8.3
Hardness of Materials

Aluminum	2–2.9
Brass	3–4
Copper	2.5–3
Diamond	10
Gold	2.5–3
Iron	4–5
Lead	1.5
Magnesium	2.0
Marble	3–4
Mica	2.8
Phosphorbronze	4
Platinum	4.3
Quartz	7
Silver	2.5–4
Steel	5–8.5
Tin	1.5–1.8
Zinc	2.5

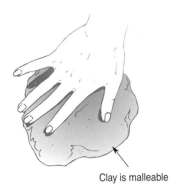

Clay is malleable

Figure 8.18
The malleability of a material is a measure of how easily it can change shape.

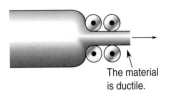

The material is ductile.

Figure 8.19
Materials that can easily be extruded through a die are ductile.

 ## DUCTILITY

Ductile materials are easily deformed when under tension, such as when metal is extruded through a die to make a wire (Fig. 8.19).

Ductile materials have a large plastic range under tension. Steel is both malleable and ductile when heated above 1,400°F.

CHAPTER SUMMARY

$$F = k \cdot \Delta x$$

Force = spring constant × length change

$$Stress = \frac{F}{A}$$

$$Stress = \frac{force}{area}$$

$$Strain = \frac{\Delta L}{L}$$

$$Strain = \frac{length\ change}{original\ length}$$

$$Y = \frac{F/A}{\Delta L/L}$$

$$Young's\ modulus = \frac{stress}{strain}$$

$$B = -\frac{F/A}{\Delta V/V}$$

$$Bulk\ modulus = -\frac{volume\ stress}{volume\ strain}$$

$$S = \frac{F/A}{d/l}$$

$$Shear\ modulus = \frac{shearing\ stress}{shearing\ strain}$$

Problem-Solving Tip

■ When solving problems dealing with a modulus and its units, be careful to use similar units in the other variables in the equation.

PROBLEMS

1. Suppose a man who weighs 165 Lb gets in his car, and the car lowers by 0.10 in. What is the spring constant of the coils in the suspension?

2. If the same car in problem 1 is loaded with 552 Lb, how far down will the car lower?

3. A brick building stands 130 ft tall. Each brick weighs 10 Lb and has dimensions of 8 in by 4 in by 2 in. What is the stress on the bottom layer of bricks? How tall could a building be before the elastic limit is reached?

4. Compare the strength of a two-by-four made of pine wood measuring 1.5 in by 3.5 in and a piece of steel with the same dimensions. How much weight can they support?

5. What does the elastic limit mean?

6. What is the difference between ductility and malleability?

7. If a 560-Lb weight is hanging from a steel cable with a cross-sectional area of 0.5 in^2 and a length of 25 ft, how much will it stretch?

8. If a spring has a spring constant of 34 N/m, how far will it stretch under 2.0 N of force?

9. What is the difference between stress and strain?

10. What material would you choose to build something that could withstand a high compressive stress and still be relatively light?

11. What is the stress on your body from atmospheric pressure?

12. What is the change in the volume of a steel submarine at the bottom of the sea if the pressure at the bottom of the sea is 1,400 pounds per square inch, and the volume of the sub at the surface is 12×10^4 ft^3?

13. If a steel bolt, 0.50 in diameter, is poking out of a machine 1.0 in and has a 500.0-Lb load on it, how much does it bend?

14. How could you use a spring to measure weight? Discuss how you would calibrate it.

15. What percent will a 0.25-in diameter steel cable stretch if it is holding up 250 Lb?

Fluids

9

This chapter describes properties of fluids. Anything that flows can be considered a fluid. Even gases such as air can be treated as fluids. The principles developed in this chapter and the next will help you understand cooling electronics, working with hydraulics, aeronautics, buoyancy, pumps, compressors, and heating and ventilation systems.

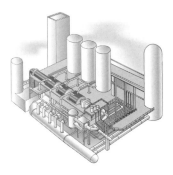

Figure 9.1

 PROPERTIES OF FLUIDS

A fluid is a material without a fixed shape. Place it in a container, however, and it will take that shape. If you squeeze some fluid out of an eyedropper, it takes a spherical shape. A droplet takes this shape because of *cohesion,* which is the attractive force between like molecules (Fig. 9.2).

Surface tension in a fluid is responsible for supporting small objects, like a waterbug, on the surface of water (Fig. 9.3).

Figure 9.2
Cohesion pulls the fluid into a spherical shape.

Figure 9.3
Surface tension in a fluid supports small objects.

Figure 9.4
Oil does not mix with water because the cohesion of the oil molecules is greater than the adhesion between the oil and water.

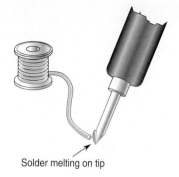

Solder melting on tip

Figure 9.5
"Wetting" a soldering iron tip. A hot iron reduces the cohesion of the molten solder until it flows over the tip.

(a)

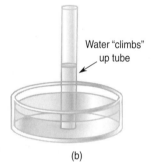

Water "climbs" up tube

(b)

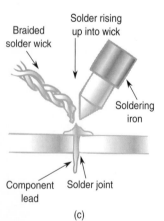

Braided solder wick

Solder rising up into wick

Soldering iron

Component lead

Solder joint

(c)

Figure 9.6
Capillary action.

Surface tension results from cohesion. It reduces water's ability to flow into small places such as fabrics. Surface tension goes down when temperature goes up. Detergents also reduce the surface tension in water. Washing your clothes in warm water with detergent reduces this tension, facilitating the water's ability to carry away the dirt.

Adhesion is the attractive force between unlike molecules. Glue and tape are considered adhesives. Whether a fluid beads up on a surface or disperses depends on the relative strength of the cohesion between the molecules of the fluid compared with the adhesion between the fluid and the surface (Figs. 9.4 and 9.5).

CAPILLARY ACTION

Capillary action is the action of fluids rising up narrow tubes and the reason that some paper towels are so absorbent. Capillary action arises from the adhesion between the fluid and the wall of the tube. Fluid will rise upward against the force of gravity until a balance is struck between the adhesion, cohesion, and gravity (Fig. 9.6).

VISCOSITY

The *viscosity* of a fluid is a measure of how easily it flows under its own weight (Fig. 9.7).

The viscosity of a fluid can be reduced with increasing temperature. Motor oils have a viscosity rating as shown in the SAE number. Oil with an SAE rating of 5 flows more easily than that with an SAE rating of 30. In cold weather, oils with low SAE numbers are used; otherwise the engine would not turn over. In warm weather, high-viscosity oils are used to prevent the oil from becoming too thin and leaking at the seals in the engine.

SPECIFIC GRAVITY

The *specific gravity* of a substance is its density divided by the density of water:

$$\text{Specific gravity} = \frac{\text{density of substance}}{\text{density of water}}$$

A substance will sink in water if its specific gravity is greater than water.
A substance will rise in water if its specific gravity is less than water.
A substance will remain still in water if its specific gravity is equal to water.

Specific gravity has no unit; it is just a number. The density of water is 1 g/cm^3. The density of lead is 11.4 g/cm^3. Therefore, the specific gravity of lead is 11.4. This is much higher than water, and therefore lead will sink in water.

Figure 9.7
Syrup is a very viscous fluid compared with water.

DETERMINING CONCENTRATIONS OF SOLUTIONS

How can you tell if you have enough antifreeze in your radiator? If the concentration is not high enough, the water may freeze! We can use the specific gravity of substances to determine it.

HYDROMETER

A hydrometer measures the specific gravity of a substance (Fig. 9.8).

The hydrometer contains a vial sealed with air and a weight, and the combination has the same density as water. As the solution is drawn into the hydrometer, the vial will float to some level where the buoyancy force balances out the weight of the vial. This level is dependent on the density of the fluid. The specific gravity is determined from a scale on the side of the hydrometer.

Antifreeze is made from ethylene glycol and can prevent the water in your car radiator from freezing, depending on the concentration in your radiator. The hydrometer can read the specific gravity of the radiator fluid and therefore determine the concentration of the antifreeze. The concentration determines the fluid's freezing temperature (see Table 9.1).

Figure 9.8
A hydrometer.

Table 9.1
Freezing Points for Various Concentrations of Antifreeze

Specific Gravity	Percent of Ethylene Glycol, by Volume	Freezing Point, °C
1.000	0	0
1.013	9.2	−3.6
1.026	18.3	−7.9
1.040	28.0	−14.0
1.053	37.8	−22.3
1.067	47.8	−33.8
1.079	58.1	−49.3
1.109	100	−17.4

A car battery contains a solution of sulfuric acid and water. If the battery is fully charged, the solution is about 38% sulfuric acid. When the battery is dead, the solution is about 26% sulfuric acid. The specific gravity changes from 1.28 fully charged to about 1.18 discharged. By measuring the specific gravity of the battery's solution, the condition of the battery can be determined.

BUOYANCY

What makes a helium balloon float in the air or a piece of wood float in the water (Fig. 9.9)?

Consider a bottle floating in water. The part of the bottle under water is displacing some volume of water. Before the bottle displaced the water, that volume of water was supported by a force equal to its weight. Otherwise, it would have sunk. This fact leads to Archimedes' principle.

Figure 9.9

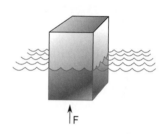

Figure 9.10

Figure 9.11

ARCHIMEDES' PRINCIPLE

The Archimedes' principle states that the upward force on a body is equal to the weight of the volume of fluid displaced by the object.

The upward force is called *buoyant force* (Fig. 9.10).

$$F_b = V \cdot D_w$$

Buoyant force = volume displaced × weight density

If you lift someone in a swimming pool, he or she feels much lighter because of Archimedes' principle (Fig. 9.11).

EXAMPLE 9.1

Suppose a 180-Lb person is partially submerged in a swimming pool. Suppose the volume of the submerged part equals 1.5 ft^3. What is the buoyant force, and how much does the person now weigh?

$$\text{Weight} = 180 \text{ Lb}$$
$$V = 1.5 \text{ ft}^3$$
$$D_w = 62.4 \text{ Lb/ft}^3$$
$$F_b = V \cdot D_w$$
$$F_b = (1.5 \text{ ft}^3)\left(62.4 \frac{\text{Lb}}{\text{ft}^3}\right) = 94 \text{ Lb}$$
$$\text{New weight} = 180 \text{ Lb} - 94 \text{ Lb} = 86 \text{ Lb}$$

No wonder we can pick up someone much bigger than us in a swimming pool!

Figure 9.12

 ## PRESSURE

Pressure is a measure of how much force is applied to a given area. Consider placing a 5-Lb weight on your arm. That might not feel like much, but what if you apply 5 Lb on top of a needle onto your arm. Ouch! That same force, applied to a small area, creates a lot of pressure. Mathematically, pressure is determined by the following equation:

$$P = \frac{F}{A}$$

Pressure = force/area

	English System	SI System
Pressure	Lb/in^2 (psi)	N/m^2 (pascals, Pa)

Conversions:
1 atmosphere = 14.7 Lb/in^2 = 101.3 kPa = 30 in of Hg

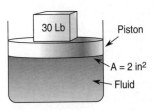

Figure 9.13

EXAMPLE 9.2

A 30.0-Lb weight pushes down on a fluid through a piston of area 2.0 in^2. What pressure is the piston applying to the fluid?

$$A = 2.0 \text{ in}^2$$
$$F = 30.0 \text{ Lb}$$
$$P = ?$$
$$P = F/A$$
$$P = 30.0 \text{ Lb}/2.0 \text{ in}^2$$
$$P = 15 \text{ Lb/in}^2$$

Figure 9.14
You cannot localize a pressure on part of a fluid.

One of the differences between a fluid and a solid is that you cannot localize a pressure on part of a fluid (Fig. 9.14). The liquid squishes out of the way.

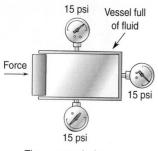

15 psi   Vessel full
of fluid

Force

15 psi

15 psi

The pressure is the same
everywhere in the
chamber, as depicted by
the pressure gauges.

Figure 9.15
Pressure applied to a
volume of fluid at the
same depth has the same
pressure throughout.

Pressure applied to a volume of fluid at the same depth has the same pressure throughout (Fig. 9.15).

 ## MEASURING PRESSURE

When you measure your tire pressure, what is it exactly that the pressure gauge is measuring?

Inside a tire there is a lot of air, which is confined to a small place. The air is composed of molecules moving around, with some striking the inside of the inner tube. The tube bulges out because it is being struck by the air molecules from the inside, forcing it outward (Fig. 9.16). If you step on the tube, the rest of the tube bulges out more because the same amount of air is confined to a smaller place, resulting in more molecules pushing or applying force to a smaller area. A pressure gauge measures how much force is applied to a given area (Fig. 9.17).

Air molecules moving
about inside a tire

Figure 9.16
Air moving around inside
a tire pushes the walls of
the tire outward.

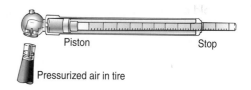

Piston   Stop

Pressurized air in tire

Figure 9.17
A typical tire gauge consists of a
piston contained in a cylinder
constrained by a spring. The piston is
connected to a calibrated rod from
which the pressure is read.

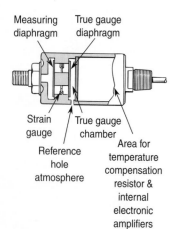

Measuring   True gauge
diaphragm   diaphragm

Strain   True gauge
gauge   chamber

Reference   Area for
hole   temperature
atmosphere   compensation
              resistor &
              internal
              electronic
              amplifiers

Figure 9.18
A strain gauge used to
measure pressure.

An electronic pressure gauge typically uses a *strain gauge,* a device that, when strained or made to change shape, will change its electrical properties (Fig. 9.18). In a pressure gauge, the strain gauge is mounted in some type of chamber that can be attached to the liquid or gas volume being measured. As the liquid or gas molecules strike the gauge, they deliver a force to the resistive pattern the sensor is made of. This will cause it to deform, resulting in a change in resistance.

A *differential pressure gauge* measures the difference in pressure between a fixed pressure and the atmosphere. Most pressure gauges measure relative to the atmospheric pressure. These relative pressures are called *gauge pressures* and are designated psig, which stands for pounds per square inch relative. Psia means pounds per square inch absolute. This is a scale relative to a perfect vacuum. See Fig. 9.19.

If a tire has a pressure of 32 psi, this refers to psig or gauge pressure.

To convert to absolute pressure, you must add the atmospheric pressure to this number.

$$\text{psia} = \text{psig} + 14.7 \text{ lb/in}^2$$

At the bottom of a swimming pool, the pressure is higher than near the top. This is because at the bottom there is a lot of water above you pushing down on you. The atmosphere itself exerts a pressure on us. There are miles of atmosphere above us, weighing us down (Fig. 9.20). Air is very light, but enough of it can weigh a substantial amount.

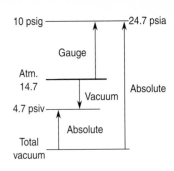

Figure 9.19
Chart comparing psia and psig.

Figure 9.20
You are essentially supporting the atmosphere above you. If you hold out your hand, 1 in^2 on your palm is supporting 14.7 Lb of air above you.

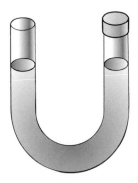

Figure 9.21
A simple barometer.

 ## BAROMETER

A *barometer* measures the pressure of the atmosphere. In a simple barometer, the height of the liquid depends on the weight of the liquid versus the force on the liquid from the atmosphere (Fig. 9.21).

 ## PRESSURE VERSUS DEPTH

Pressure in a fluid increases with depth. The pressure at the bottom of a swimming pool is higher than near the surface. The pressure does not depend on the shape of the container but only on the depth and weight density and is determined by:

$$P = h \cdot D_w$$
$$\text{Pressure} = \text{height} \times \text{weight density}$$

This equation says that the pressure in a fluid increases with depth below the surface.

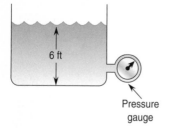

Figure 9.22
The pressure at the bottom of a pool is a combination of the pressure due to the water plus atmospheric pressure.

EXAMPLE 9.3

A swimming pool is 6 ft deep. What is the pressure at the bottom?

The pressure at the bottom of a pool is a combination of the pressure due to the water plus atmospheric pressure (Fig. 9.22).

$$h = 6.00 \text{ ft}$$

$$D_w = 62.4 \text{ Lb/ft}^3 \quad \leftarrow \quad \text{Weight density of water}$$

$$P = h \cdot D_w$$

$$P = (6.00 \cancel{\text{ft}})\left(62.4 \frac{\text{Lb}}{\text{ft}^{\cancel{3}^2}}\right)$$

$$P = 374 \frac{\text{Lb}}{\text{ft}^2}$$

Converting from $\dfrac{\text{Lb}}{\text{ft}^2}$ to $\dfrac{\text{Lb}}{\text{in}^2}$ gives

$$374 \frac{\text{Lb}}{\text{ft}^2}\left(\frac{1 \text{ ft}}{12 \text{ in}}\right)^2 = 374 \frac{\text{Lb}}{\cancel{\text{ft}^2}}\left(\frac{\cancel{\text{ft}^2}}{12^2 \text{in}^2}\right) = 2.60 \frac{\text{Lb}}{\text{in}^2}$$

$$P = 2.60 \frac{\text{Lb}}{\text{in}^2} \text{ psig} \quad \text{gauge pressure}$$

The pressure at the bottom of the pool is $2.60 \dfrac{\text{Lb}}{\text{in}^2}$ greater than at the surface.

$$\text{Absolute pressure} = 2.60 \frac{\text{Lb}}{\text{in}^2} + 14.7 \frac{\text{Lb}}{\text{in}^2}$$

$$\text{Absolute pressure} = 17.3 \frac{\text{Lb}}{\text{in}^2}$$

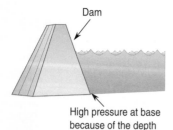

Dam

High pressure at base because of the depth

Figure 9.23
Pressure at the bottom of a dam is much greater than at the top. Therefore, the dam is built much wider near the bottom to support the extra pressure.

The reason a dam is built with a much wider bottom is because of this increasing pressure with depth. See Fig. 9.23.

 HYDRAULICS

Hydraulics is about transferring pressure from one region to another to obtain some mechanical advantage. A hydraulic jack can generate a force much greater than an applied force to the jack handle (Fig. 9.24). Inside the jack there are two chambers with pistons. The jack handle is connected to a small piston in the small chamber that pressurizes the fluid. When the jack handle is pushed down with some force, a pressure in the fluid builds up and is transferred to the big piston, creating a much greater force upward than the applied force. One-way valves prevent the fluid from moving in the wrong direction.

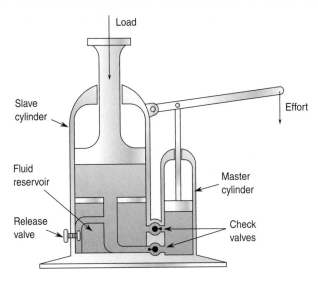

Figure 9.24
A hydraulic jack.

In a hydraulic jack, the pressures inside the chambers are equal. The pressure is generated by the small piston and transferred through the fluid to the big piston. Because the pressures are equal:

$$P_{in} = P_{out}$$

$$\frac{F_{in}}{A_{in}} = \frac{F_{out}}{A_{out}}$$

Therefore

$$F_{out} = \left(\frac{A_{out}}{A_{in}}\right) \cdot F_{in}$$

This equation says the force going in can be increased if the area of the out-piston is bigger than the in-piston.

A hydraulic jack is a machine and therefore obeys the law of machines:

$$F_{in} \cdot d_{in} = F_{out} \cdot d_{out}$$

It therefore follows that

$$\frac{F_{out}}{F_{in}} = \frac{A_{out}}{A_{in}} = \frac{d_{in}}{d_{out}}$$

This equation says that the small piston will travel through a greater distance than the big piston if the output force is larger than the input force. In other words, you have to pump the jack handle up and down many times to lift a car just a little.

EXAMPLE 9.4

A hydraulic press contains a small and a large piston of areas 3 in² and 30 in², respectively. The handle of the jack to the small piston delivers 50 Lb of force. How much force is generated?

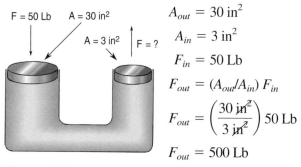

$$A_{out} = 30 \text{ in}^2$$

$$A_{in} = 3 \text{ in}^2$$

$$F_{in} = 50 \text{ Lb}$$

$$F_{out} = (A_{out}/A_{in}) \, F_{in}$$

$$F_{out} = \left(\frac{30 \text{ in}^2}{3 \text{ in}^2}\right) 50 \text{ Lb}$$

$$F_{out} = 500 \text{ Lb}$$

Figure 9.25

BRAKING SYSTEMS

Automotive brakes work on a hydraulic system (Fig. 9.26). They consist of a master cylinder, which is a reservoir of fluid, and a brake cylinder at each wheel. The master cylinder contains two pistons, one for the front and one for the back brakes. When the brake pedal is pressed, the pressure generated by the master cylinder is transmitted to each wheel cylinder, causing the brake shoes or the calipers to apply a mechanical pressure to the drums, or calipers, to slow the vehicle down.

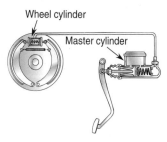

Figure 9.26
A hydraulic brake system.

Hydraulic systems are also used in heavy equipment. For this use, a pump generates the pressure in the master cylinder.

PNEUMATICS

Pneumatics is the fluid mechanics of air. Air is considered a low-viscosity fluid. One application of pneumatics is the air hammer, which consists of a heavy piston that oscillates up and down with air pressure controlled by a rocker valve (Fig. 9.27).

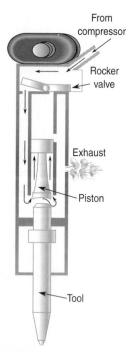

Figure 9.27
An air hammer.

CHAPTER SUMMARY

Cohesion is the attractive force between like molecules.

Adhesion is the attractive force between unlike molecules.

Capillary action is the behavior of fluids as they rise up narrow tubes and the reason that some paper towels are so absorbent.

Viscosity of a fluid is a measure of how easily it flows under its own weight.

Pressure = force/area

$$P = \frac{F}{A}$$

$$P = h \cdot D_w$$

Pressure = height × weight density

Archimedes' principle states that the upward force on a body is equal to the weight of the volume of fluid displaced by the object.

$$F_b = V \cdot D_w$$

Buoyant force = volume displaced × weight density

The **specific gravity** of a substance is its density divided by the density of water.

$$\text{Specific gravity} = \frac{\text{density of substance}}{\text{density of water}}$$

For a hydraulic jack:

$$\frac{F_{out}}{F_{in}} = \frac{A_{out}}{A_{in}} = \frac{d_{in}}{d_{out}}$$

Problem-Solving Tips

- Be careful to monitor your units, making sure you are in the same system of units throughout each problem.

PROBLEMS

1. What is the difference between cohesion and adhesion?

2. Explain capillary action.

3. Why is some motor oil multigrade?

4. If pressure is responsible for moving fluids, can you explain why a low barometric pressure reading indicates a storm is approaching?

5. What is the difference between psig and psia?

6. What is the pressure relative to the atmosphere at the bottom of a swimming pool that is 6 ft deep?

7. If a ship displaces 25,000 ft^3 of water, what is the buoyant force?

8. Given an air bag and an air compressor, how could you raise a sunken ship?

9. Design a hydraulic jack that is able to lift 1,500 Lb with 50 Lb of force. What does the ratio of the pistons have to be?

10. What is the purpose of the oil in a hydraulic jack?

11. Why does a hot air balloon float?

12. If a rectangular piece of wood measuring 2.5 in deep by 3.5 in wide by 6.0 in long is thrown into water, how much of the wood will be submerged?

13. If the specific gravity of your radiator fluid is 1.053, what temperature will it freeze at?

14. How does solder wick remove solder?

15. How does detergent aid in cleaning your clothes?

16. Suppose a pressure gauge uses a strain gauge that changes resistance with pressure as $125 \frac{\Omega}{\text{psi}}$. What is the pressure change if the resistance of the gauge changes by 290 Ω?

17. If a flow sensor monitors the flow through a pipe and the viscosity of the fluid decreases, what happens to flow rate?

10

Fluid Flow

This chapter describes fluid flow and its applications. Understanding fluid flow is important in the electronics field. Air can be considered a fluid, and its flow around electrical equipment is vital for cooling. Knowledge of this subject should also help you understand plumbing, pneumatics, aerodynamics, and ventilation.

Figure 10.1

TYPES OF FLUID FLOW

Basically, there are two types of fluid flow: laminar and turbulent flow (Fig. 10.2). *Laminar flow* is smooth, streamlined flow. *Turbulent flow* is nonlaminar flow, characterized by whirlpools and eddies.

The reduction of turbulent flow is of major concern to automobile and aeronautical engineers (Fig. 10.3). Fluid flow is impeded by friction. *Friction* in a

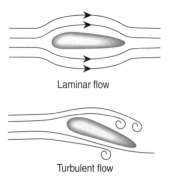

Laminar flow

Turbulent flow

Figure 10.2
Laminar and turbulent flow

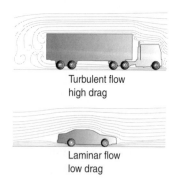

Turbulent flow
high drag

Laminar flow
low drag

Figure 10.3
Turbulent flow can be minimized by changing the shape of a vehicle.

flow is determined by the viscosity of the fluid and the type of flow, that is, whether it is streamlined or turbulent. Creating a smooth surface for the fluid to flow over can reduce friction (Fig. 10.4). The drag on a moving vehicle is high when there is turbulent flow. Turbulent flow wastes energy, and minimizing it is almost always desired.

Aerodynamic drag on an automobile is small at low velocities, but it is responsible for most of the reduction in fuel economy at speeds over 40 mph. The aerodynamic drag is given by the following:

$$F_{Drag} = \frac{C_D \cdot A_f \cdot v^2}{370} \quad \leftarrow \quad F_{Drag} \text{ is in Lb, } A \text{ is in ft}^2, \text{ and } v \text{ is in mph.}$$

$$\text{Drag force} = \frac{\text{aerodynamic drag coefficient} \times \text{frontal area} \times \text{velocity}^2}{370}$$

The drag force is measured in pounds. The aerodynamic drag coefficient is a measure of how streamlined the car is. Frontal area in square feet is the total area facing forward, including mirrors, luggage rack, and so on, and the velocity is expressed in miles per hour.

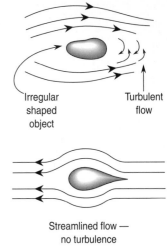

Figure 10.4
A teardrop has the lowest drag of any shape for speeds below the speed of sound.

Table 10.1
Aerodynamic Drag Coefficients

Square flat plate (worst)	1.17
Rectangular block	1.00
Ordinary truck	0.70
1972 Dodge Polara Wagon	0.60
Streamlined truck	0.55
1981 Cadillac Eldorado	0.55
Porsche 928	0.45
Jaguar XKE	0.40
Ford Escort	0.39
Camaro	0.39
Datsun 280 ZX	0.39
Porsche 924	0.34
1992 Ford Taurus	0.32
1996 Ethos electric car	0.31
1997 Audi A8	0.29
Lotus Europa	0.29
VW research vehicle	0.15
Teardrop (theoretical best)	0.03

EXAMPLE 10.1

Suppose you put your hand, with an area of 0.15 ft², out a car window moving at 65 mph. What is the force on your hand?

$$v = 65 \text{ mph}$$

$$A_f = 0.15 \text{ ft}^2$$

Let us approximate your hand as a square flat plate having a $C_D = 1.17$

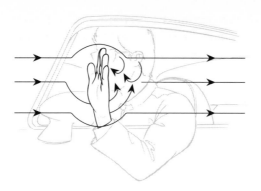

Figure 10.5

$$F_D = ?$$

$$F_{Drag} = \frac{C_D \cdot A_f \cdot v^2}{370}$$

$$F_{Drag} = \frac{1.17 \cdot 0.15 \cdot 65^2}{370}$$

$$F_{Drag} = 2.0 \text{ Lb}$$

Turbulent flow can be used to your advantage when keeping a car firmly on the road at high speeds. A spoiler changes the laminar flow over the car to turbulent flow to keep the car from lifting off the road like an airplane (Fig. 10.6).

Figure 10.6
A spoiler.

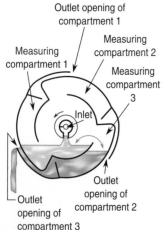

Outlet opening of compartment 1

Measuring compartment 2

Measuring compartment 1

Measuring compartment 3

Inlet

Outlet opening of compartment 2

Outlet opening of compartment 3

Figure 10.7
Volumetric flow meter.

MEASURING THE VOLUME OF A FLUID IN A FLOW

In order to charge you properly, the gas pump must measure how much gasoline you use. Likewise, the utility company must measure how much water you use. To measure the volume of a fluid, a volumetric meter is used (Fig. 10.7). As the water enters the meter, it fills chambers that have a known volume. As the chambers fill, they rotate, emptying their contents into the output. A recording mechanism records how many revolutions the chambers have made.

Another type of volumetric flow meter commonly used in metering household water consumption uses a rotating impeller (Fig. 10.8). As the fluid flows through the meter, the impeller rotates. The number of rotations is proportional to the amount of fluid passing through the meter. A counting mechanism is mounted on the rotating shaft, indicating the volume passed.

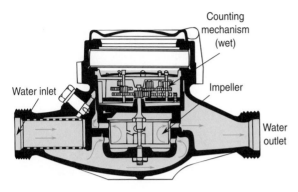

Figure 10.8
Impeller-type volumetric fluid meter.

 ## MEASURING FLOW RATE

Fans are used to keep electronics cool by sustaining a proper flow rate. Flow rates are measured in cubic feet per minute (cfm) or gallons per minute (gpm) or some other volume per time unit. Flow rate can be measured using a tachometer attached to a propeller turning in the flow (Fig. 10.9) or a thermal sensor (Fig. 10.10).

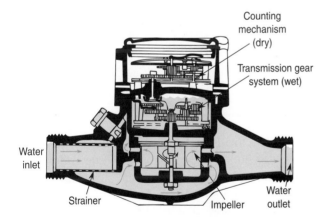

Figure 10.9
A tachometer attached to an impeller-type flow meter indicates the rate of volume of fluid flowing through the meter.

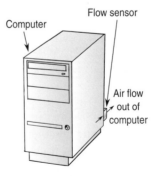

Figure 10.10
A thermal flow sensor measuring the air flow through a computer.

A thermal flow sensor contains a heater and an electronic thermometer placed next to one another. As the fluid flows over the heater, the heater cools. The greater the flow, the cooler the heater will be. The temperature is therefore a measure of the flow rate.

 ## CALCULATING FLOW RATE

The rate at which a volume fluid flows through a pipe is given in terms of

$$\frac{\text{gallons}}{\text{minute}}, \frac{\text{cm}^3}{\text{min}}, \text{ or any other } \frac{\text{volume}}{\text{time}} \text{ unit}$$

The rate of flow is given by:

$$R = A \cdot v$$

Rate of flow = Area of pipe × velocity of fluid

This equation says that a large pipe with fluid flowing quickly through it will have a high flow rate.

EXAMPLE 10.2

A pipe with a cross-sectional area of 1.25 cm² has water flowing through it at 10.0 cm/s. What is the flow rate?

$$A = 1.25 \text{ cm}^2$$

$$v = 10.0 \frac{\text{cm}}{\text{s}}$$

$$R = ?$$

$$R = A \cdot v$$

$$R = (1.25 \text{ cm}^2)\left(10.0 \frac{\text{cm}}{\text{s}}\right)$$

$$R = 12.5 \frac{\text{cm}^3}{\text{s}}$$

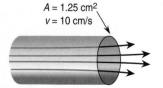

$A = 1.25 \text{ cm}^2$
$v = 10 \text{ cm/s}$

Figure 10.11

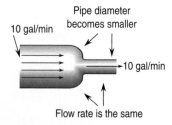

10 gal/min

Pipe diameter becomes smaller

10 gal/min

Flow rate is the same

Figure 10.12
The rate at which an amount of fluid flows through a pipe does not change even if the pipe size changes.

Suppose a fluid flows through a pipe, and the diameter of the pipe decreases. If the fluid is incompressible (not compressible under pressure), then the rate of fluid that flows into the big pipe must equal the rate of flow out of the small pipe (Fig. 10.12).

As a fluid flows through an area, if the area gets smaller, the fluid speeds up. If the area becomes larger, the fluid moves more slowly. This is similar to a deep, slow-moving river turning into shallow, quick-moving rapids. This is summarized in what is called the *continuity equation:*

$$A_1 \cdot v_1 = A_2 \cdot v_2$$

Area 1 × velocity 1 = area 2 × velocity 2

EXAMPLE 10.3

Suppose water is flowing through a pipe with a cross-sectional area of 1.5 in^2 and a velocity of 2.0 in/s. If the pipe area decreases to 1.0 in^2, what is the new velocity of the water?

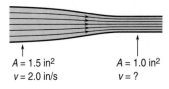

$$A_1 = 1.5 \text{ in}^2$$

$$v_1 = 2.0 \text{ in/s}$$

$$A_2 = 1.0 \text{ in}^2$$

$$v_2 = ?$$

$$A_1 \cdot v_1 = A_2 \cdot v_2$$

$$v_2 = \frac{A_1 \cdot v_1}{A_2}$$

$$v_2 = \frac{(1.5 \text{ in}^2)(2.0 \text{ in/s})}{1.0 \text{ in}^2}$$

$$v_2 = 3.0 \text{ in/s}$$

Figure 10.13

 FLUID PUMPS

Fluid flows when there is a pressure difference, and pumps are used to generate this pressure. Pumps are rated in terms of their flow rate, volume/time, and the pressure they generate.

Piston Pumps

On the first stroke, the piston moves inward, creating a partial vacuum, which allows the atmospheric pressure to force the fluid into the inlet of the pump. On the second stroke, the piston pushes the fluid out (Fig. 10.14). Inlet and outlet valves ensure the fluid moves only in the desired direction (Fig. 10.15).

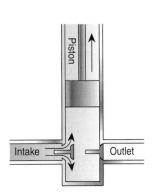

Figure 10.14
A piston pump alternately
draws fluid in, then out,
with each stroke.

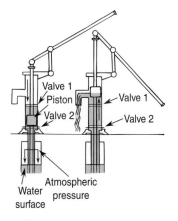

Figure 10.15
A well pump is a type of
piston pump.

Centrifugal Pump

In a centrifugal pump, a rotating impeller deflects the incoming fluid along the axis of the impeller to the outlet port (Fig. 10.16). A squirrel cage blower is a type of centrifugal pump (Fig. 10.17). The flow outward is continuous, unlike with the piston pump.

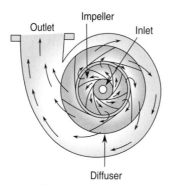

Figure 10.16
A centrifugal pump.

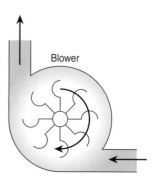

Figure 10.17
A squirrel cage blower for air is an example of a centrifugal pump.

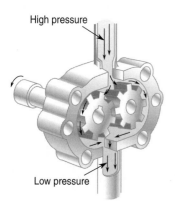

Figure 10.18
A gear pump.

Gear Pump

A gear pump drives the fluid around the outside of the gears toward the outlet in a continuous manner (Fig. 10.18).

PUMP POWER

The power of a pump needed to deliver a fluid at a given flow rate and pressure can be calculated by:

$$hp = \frac{gpm \times psi}{1,714}$$

$$Horsepower = \frac{gallons\ per\ minute \times pounds\ per\ square\ inch}{1,714}$$

EXAMPLE 10.4

A fountain pump is needed to pump 12 gpm of water at a pressure of 24 psi. What horsepower does the pump need to be?

$$gpm = 12$$
$$psi = 24$$
$$hp = ?$$

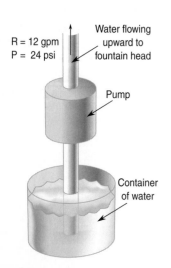

R = 12 gpm
P = 24 psi

Water flowing upward to fountain head

Pump

Container of water

Figure 10.19

$$hp = \frac{gpm \times psi}{1,714}$$

$$hp = \frac{12 \times 24}{1,714}$$

$$hp = 0.17 \text{ hp}$$

Static

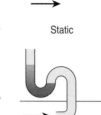

Velocity

Vacuum

Figure 10.20
Three types of pressure in a moving fluid: static pressure, velocity pressure, and vacuum pressure.

PRESSURE IN A MOVING FLUID

Pressure drives a fluid through a pipe. The higher the pressure across the fluid, the greater the speed. There are three types of pressure to consider: static pressure, velocity pressure, and vacuum pressure (Fig. 10.20).

Note that in Figure 10.20 the static pressure drops where the fluid velocity is highest. This pressure change is a result of *Bernoulli's principle,* which states that in the absence of friction, the energy of the fluid is constant throughout the pipe. The total energy of the fluid is:

$$E_{Total} = \text{kinetic energy} + \text{energy from pressure} + \text{gravitational potential energy}$$

So why would the pressure be lowest where the fluid velocity is highest? As the fluid enters the smaller section of pipe, it speeds up. Therefore, its kinetic energy goes up. If the total energy is constant, then one of the other energy terms must decrease, that is, the pressure term. The fluid speeds up at an energy cost to the energy in the pressure term. You may have experienced this in the shower. The water flow reduces the pressure in the shower (Fig. 10.21).

The effect of low pressure created by a rapidly moving fluid is partially responsible for how an airplane gets its lift. The plane's wing is shaped so that the air moving over the top of the wing has to move faster to keep up with the air moving below it. As a result, the pressure is higher below the wing than above it, and the difference in air pressure pushes the plane upward (Figs. 10.22 and 10.23).

Shower curtain moves inward toward low pressure

Figure 10.21
Low pressure created by a rapidly moving fluid.

Low pressure

High pressure

Figure 10.22
When the wing is horizontal, air moving over the top has to move faster to keep up with the air moving below, resulting in a pressure difference.

Figure 10.23
The plane also receives a lift when the wing is tilted slightly upward. In this case, air is deflected off the bottom of the wing, pushing it upward.

CARBURETOR

An automobile's carburetor contains a venturi, or a narrowing in the throat of the carburetor, where the gas and air mix (Fig. 10.24). In this region, air flows quickly, reducing the pressure there. As a result of this lower pressure, gas flows into the venturi from a higher-pressure bowl of gas. A needle valve controls the amount of gas in the bowl. The lower pressure is responsible not only for the gas flow into the venturi but also for mixing the gas and air more thoroughly because the gas evaporates more easily at lower pressures.

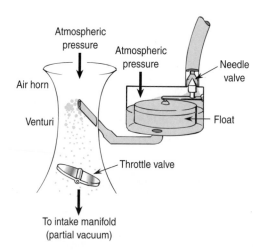

Figure 10.24
The carburetor.

POWER GENERATION

Kinetic energy, captured from the wind and flowing water, can be turned into electrical energy.

Wind Power

A windmill uses kinetic energy from wind for pumping water, milling corn, generating electricity, and other purposes. How much energy does wind have? The amount depends on the wind's density and moisture content. Under average conditions, the power contained in a cross-sectional area (A) of wind is:

$$P = 0.61 \cdot v^3 \cdot A \quad \leftarrow \quad P \text{ is in W, } v \text{ is in m/s, and } A \text{ is in m}^2.$$

Power $= 0.61 \cdot (\text{velocity})^3 \cdot \text{area}$

The power is proportional to the velocity of the wind cubed.

EXAMPLE 10.5

Figure 10.25

Suppose the blades of a windmill can effectively capture an area of wind 8.5 m^2 in size and the wind is blowing at 4.0 m/s. How much power can be captured?

$$v = 4.0 \text{ m/s}$$
$$A = 8.5 \text{ m}^2$$
$$P = 0.61 \cdot v^3 \cdot A$$
$$P = 0.61 \cdot (4.0 \text{ m/s})^3 \cdot 8.5 \text{ m}^2$$
$$P = 330 \text{ W}$$

Hydroelectric Power

Hydroelectric power is power generated from moving water. How much energy is contained in moving water?

$$P = 500 \cdot v^3 \cdot A \quad \leftarrow \quad P \text{ is in W, } v \text{ is in m/s, and } A \text{ is in m}^2.$$
$$\text{Power} = 500 \cdot (\text{velocity})^3 \cdot \text{area}$$

This equation is almost identical to the one for wind power. They are different because the density of water is much greater than the density of air.

EXAMPLE 10.6

A hydroelectric power plant consists of a dam with water turbines that spin electrical generators. Suppose the water is flowing at 15 m/s and is spinning a turbine with an effective area of 12 m². How much power is being captured?

$$v = 15 \text{ m/s}$$
$$A = 12 \text{ m}^2$$
$$P = 500 \cdot v^3 \cdot A$$
$$P = 500 \cdot (15 \text{ m/s})^3 \cdot 12 \text{ m}^2$$
$$P = 2\overline{0} \text{ MW}$$

Figure 10.26

CHAPTER SUMMARY

Types of fluid flow: There are two basic types of fluid flow, laminar and turbulent flow.

Laminar flow: Laminar flow means smooth, streamlined flow.

Turbulent flow: Turbulent flow is nonlaminar flow characterized by whirlpools and eddies.

Aerodynamic drag is given by:

$$F_{Drag} = \frac{C_D \cdot A_f \cdot v^2}{370} \quad \leftarrow \quad \begin{array}{l} F_{Drag} \text{ is in} \\ \text{Lb, } A \text{ is in ft}^2, \\ \text{and } v \text{ is in mph.} \end{array}$$

$$\text{Drag force} = \frac{\text{aerodynamic drag coefficient} \times \text{frontal area} \times \text{velocity}^2}{370}$$

The rate of flow is given by:

$$R = A \cdot v$$

Rate of flow = area of pipe × velocity of fluid

The continuity equation:

$$A_1 \cdot v_1 = A_2 \cdot v_2$$

Area 1 × velocity 1 = area 2 × velocity 2

The power of a pump needed to deliver a fluid at a given flow rate and pressure can be calculated by:

$$\text{hp} = \frac{\text{gpm} \times \text{psi}}{1{,}714}$$

$$\text{Horsepower} = \frac{\text{gallon per minute} \times \text{pounds per square inch}}{1{,}714}$$

Bernoulli's principle states that in the absence of friction, the energy of the fluid is constant throughout the pipe. The total energy of the fluid is:

$$E_{Total} = \text{kinetic energy} + \text{energy from pressure} + \text{gravitational potential energy}$$

A windmill draws power from the wind. Under average conditions, the power contained in a cross-sectional area *(A)* of wind is:

$$P = 0.61 \cdot v^3 \cdot A \quad \leftarrow \quad P \text{ is in W, } v \text{ is in m/s, and } A \text{ is in m}^2.$$

$$\text{Power} = 0.61 \cdot (\text{velocity})^3 \cdot \text{area}$$

Hydroelectric power is power generated from moving water. The amount of energy contained in moving water is given by:

$$P = 500 \cdot v^3 \cdot A \quad \leftarrow \quad P \text{ is in W, } v \text{ is in m/s, and } A \text{ is in m}^2.$$

$$\text{Power} = 500 \cdot (\text{velocity})^3 \cdot \text{area}$$

Problem-Solving Tips

- In this chapter there are a number of formulas that work only with a specific set of units, such as the fps system. Be mindful of your units!
- There are three formulas for power in this chapter. Choose the correct formula according to the context of the problem.

PROBLEMS

1. Water is flowing through a pipe of radius 0.25 in at a velocity of 2.0 ft/s. What is the rate of flow?

2. A fluid is flowing through a pipe of area 0.375 in^2 at a velocity of 3.0 ft/s. It enters the pipe with a cross section of 0.185 in^2. What is the velocity of the fluid in the smaller section of pipe? What is the rate of flow in both sides?

3. Suppose you build a windmill and connect it to a car alternator to supply power to your home. If the wind is blowing at 12.5 m/s, the area of the blades takes up 3.5 m^2, and the efficiency of the whole system is 21%, how much power can be generated?

4. Without using a pump, how can you increase the speed of a moving fluid?

5. If the power given to a hydroelectric generator by falling water is 12 MW and the turbines have an effective area of 5.5 m^2, how fast is the water moving?

6. What is the difference between laminar flow and turbulent flow?

7. How many more times is the air drag on a car moving at 25 mph than at 50 mph?

8. A pump is to deliver a flow rate of 1,200 gpm at 33 psi. What is its horsepower?

9. How much power can a windmill generate with an effective gathering area of 4.5 m^2 and an efficiency of 35% for a wind moving at 95 m/s?

10. How would you measure the flow rate and volume of a leaky faucet?

11. What shape should the auto industry make cars to minimize air drag?

12. If you ride with your foot out of a car window while traveling at 55 mph, what is the estimated force on your foot?

13. A fan is mounted over a CPU and blows at a rate of 10 cfm through a cross-sectional area of 2 ft^2. How fast is the air moving?

14. How could you measure the flow rate through a piece of electronics equipment to ensure it is being cooled properly?

15. What are the two ways an airplane gets its lift?

Temperature and Heat

This chapter discusses temperature and heat. Many problems can occur in electronics if proper temperatures are not maintained. This chapter will describe how temperature is measured and how to calculate and control it.

Figure 11.1

Exactly what is a thermometer measuring when it measures temperature? When water is heated, its molecules or atoms begin vibrating or moving about faster, increasing their kinetic energy (Fig. 11.2). This energy is transferred to the thermometer. The thermometer is therefore measuring the average kinetic energy of the water.

Figure 11.2
When a pot of water is boiled, heat is applied to the pot. The heat causes the water molecules in the pot to move faster.

 ## TEMPERATURE SCALES

There are many temperature scales. We will discuss three of them: Celcius, Fahrenheit, and Kelvin. The Celcius and Fahrenheit scales are defined in an arbitrary way. For example, in the Celcius scale, 0° and 100° are arbitrarily placed at the freezing and boiling points of water.

If temperature is a measure of moving energy, then a substance that is completely still should have a temperature of zero. The Kelvin temperature scale is based on this idea, with 0° set to be the lowest possible temperature.

Let's compare a few scales.

	Fahrenheit	Celsius	Kelvin
Freezing water	32°F	0°C	273°K
Boiling water	212°F	100°C	373°K

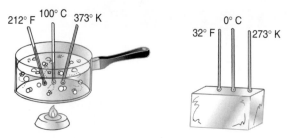

Figure 11.3
A comparison of temperature scales.

To convert between these scales, apply the following formulas:

$$°C = \frac{5}{9}(°F - 32)$$

$$°F = \frac{9}{5}°C + 32$$

$$°C = °K - 273$$
$$°K = °C + 273$$

EXAMPLE 11.1

Convert 72°F into Celsius.

$$°F = 72°F$$

$$°C = \frac{5}{9}(°F - 32)$$

$$°C = \frac{5}{9}(72°F - 32) = 22°C$$

EXAMPLE 11.2

Convert 39°C into Fahrenheit.

$$°C = 39°C$$

$$°F = \frac{9}{5}°C + 32$$

$$°F = \frac{9}{5} \cdot 39°C + 32 = 102°F$$

EXAMPLE 11.3

Convert 34°C into kelvin.

$$°C = 34.0°C$$

$$°K = °C + 273$$

$$°K = 34°C + 273 = 307°K$$

MEASURING TEMPERATURE

Glass Thermometer

A *glass thermometer* has a liquid inside it that expands in a definite way when heated. When a glass thermometer is placed into a pot of boiling water, the molecules of the liquid inside the thermometer will begin to move faster, causing the liquid to expand and rise up in the thermometer (Fig. 11.4). How much it rises determines the temperature.

Thermocouple

A *thermocouple* consists of two types of wire that are connected at one end. A voltage is developed across the junction and is proportional to the temperature of the junction. A reference junction is used to obtain measurements with respect to a known reference temperature (Fig. 11.5).

Figure 11.4
Liquid in a thermometer rises because the heat causes the liquid molecules to move faster and therefore expand.

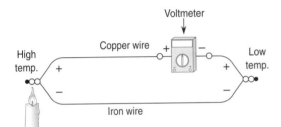

Figure 11.5
A thermocouple consists of two dissimilar pieces of wire connected to each other, producing a voltage proportional to the temperature. A reference junction scales the output to a known reference temperature.

Thermocouples can be found in hot water heaters. They are used as a safety device to monitor the pilot light. If for some reason the pilot light goes out, the gas should

Table 11.1
Thermocouple Types and Their Ranges

Two Metals	Temperature Range (°F)	Voltage Generated for Temperature Range (mV)
Iron vs. Copper–Nickel	32 to 1382	0 to 42.3
Nickel–Chromium vs. Nickel–Aluminum	−328 to 2282	−5.9 to 50.6
Platinum–6% Rhodium vs. Platinum–30% Rhodium	32 to 3092	0 to 12.4

be shut off to keep the house from filling up with gas. If the pilot light goes out, the thermocouple will alert the gas valve to shut off.

Thermistor

A thermistor is a semiconductor device that changes its resistance with temperature. As the temperature goes up, its resistance goes down (Fig. 11.6).

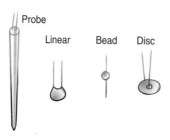

Figure 11.6
A thermistor is a variable resistor that changes resistance with temperature.

Table 11.2
Calibration for a Thermistor

°C	Ohms
−50	100,000
0	7,500
+50	7,400
+100	100
+150	50
+200	27
+250	10
+300	7.5

Bimetal Strip

Most home thermostats contain a coil consisting of two types of metal layered one on top of the other (Fig. 11.7).

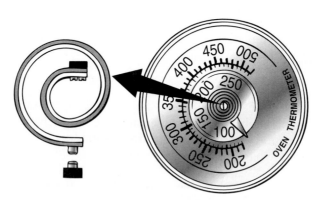

Figure 11.7
A bimetal strip thermostat.

As the temperature changes, the two metals will expand different amounts, causing the strip to bend. The strip is used as a switch to open or close a circuit, telling the furnace or air conditioner to turn on or off.

Optical Pyrometer

Optical pyrometers determine temperature by measuring the light emitted by a hot object (Fig. 11.8). For instance, as a metal is heated, it changes color from red to

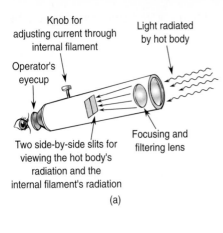

Knob for adjusting current through internal filament

Light radiated by hot body

Operator's eyecup

Two side-by-side slits for viewing the hot body's radiation and the internal filament's radiation

Focusing and filtering lens

(a)

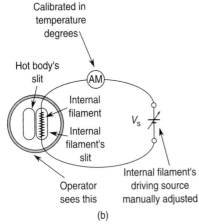

Calibrated in temperature degrees

Hot body's slit

Internal filament

Internal filament's slit

V_s

Operator sees this

Internal filament's driving source manually adjusted

(b)

Figure 11.8
An optical pyrometer determines temperature by comparing the color of an internal filament to the color of the radiation emitted by an object whose temperature is being measured.

white. Each shade of color corresponds to a particular temperature. The pyrometer converts the color it sees into a temperature.

 HEAT

In everyday language, people use the words *heat* and *temperature* to mean the same thing, but in science they are not the same. So exactly what is heat?

Heat is the energy transferred between two systems due to a temperature difference. The temperature differences between two things determine which direction, and at what rate, heat flows.

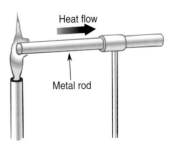

Heat flow

Metal rod

If something gets hot, it is because heat flows into it.

If something gets cold, it is because heat flows out of it.

Heat flows from hot to cold (Fig. 11.9).

Figure 11.9
Heat flows from hot to cold.

If something feels hot to the touch, that means heat is flowing into your hand. If something feels cold, it is because heat is leaving your body.

Figure 11.10
A hot cup of coffee emits heat to the room, which is at a lower temperature.

How does a thermos bottle know to keep hot coffee hot and cold water cold? The thermos bottle doesn't know. It just prevents the flow of heat (Fig. 11.10).

Heat has the same units as energy, such as joules or foot-pounds. A common unit used is the *British thermal unit* (Btu). It takes 1 Btu of heat to raise the temperature of 1 Lb of water 1°F.

Heat units and conversions:

1 Btu = 1,055 J

1 Btu = 778 ft Lb

1ft Lb = 1.36 J

1 cal = 4.186 J

1 food calorie = 1 Cal ← Note the capital C

1 Cal = 4,186 J

We can build a heat source by connecting a battery to a resistor (Fig. 11.11). Suppose the power dissipated by the circuit equals 100.0 W, and we let the circuit run for 50.0 seconds. How much heat does the resistor emit?

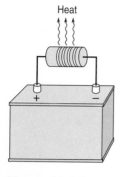

Figure 11.11

$$P = 100 \text{ W}$$

$$t = 50 \text{ s}$$

$$P = \frac{E}{t} \quad \leftarrow \quad \text{formula for power}$$

$$P \cdot t = \frac{E}{\cancel{t}} \cdot \cancel{t} \quad \text{solving for E}$$

$$E = P \cdot t$$

$$E = (100 \text{ W})(50 \text{ s})$$

$$E = 5000 \text{ J} \quad \text{of heat coming out of the resistor}$$

Converting to Btus:

$$5,000 \, \cancel{J}\left(\frac{1 \text{ Btu}}{1,055 \, \cancel{J}}\right) = 4.7 \text{ Btu}$$

This amount of energy could raise the temperature of 1 Lb of water 4.7°F.

HEAT TRANSFER

Heat transferred into finger by conduction

There are three ways to transfer heat from one place to another: conduction, convection, and radiation.

Conduction

Figure 11.12
Heat transfer by conduction, two materials that are physically touching.

Conduction is heat transferred between two materials in contact (Fig. 11.12).

How easily heat flows through a material depends on the following:

■ Heat flows because of a temperature difference across a material. The bigger the temperature differences, the larger the heat flow.

- Some materials, such as metals, allow heat to flow easily. Other materials, such as Styrofoam, prevent the flow of heat. How well they allow heat to flow is determined by their thermal conductivity *(k)*. Thermal conductivity is similar to electrical conductivity, but it is applied to heat, not electricity.

- A material that is short and has a large cross-sectional area can pass a lot more heat in a given time than a long, skinny piece of material.

These points can be summarized by the following formula:

$$Q = \frac{k \cdot A \cdot t \cdot (T_2 - T_1)}{l}$$

$$\text{Heat} = \frac{\text{thermal conductivity} \cdot \text{area} \cdot \text{time} \cdot (\text{temperature}_2 - \text{temperature}_1)}{\text{length}}$$

EXAMPLE 11.4

The handle of a cast-iron skillet has a length of 0.5 ft, a cross-sectional area of 0.0075 ft², a thermal conductivity of $k = 26\dfrac{\text{Btu}}{\text{ft°Fh}}$, and a temperature of 450°F at one end and 72°F at the other end. If we wait 0.3 hours for the heat to flow, how much heat has passed through the handle?

72° F

Q

$A = 0.0075 \text{ ft2}$

0.5 ft

450° F

Figure 11.13

$$k = 26\frac{\text{Btu}}{\text{ft°Fh}}$$

$$A = 0.0075 \text{ ft}^2$$

$$l = 0.5 \text{ ft}$$

$$t = 0.3 \text{ h}$$

$$T_2 = 450°\text{F}$$

$$T_1 = 72°\text{F}$$

$$Q = ?$$

$$Q = \frac{k \cdot A \cdot t \cdot (T_2 - T_1)}{l}$$

$$Q = \frac{26\dfrac{\text{Btu}}{\cancel{\text{ft°Fh}}} \cdot 0.0075 \cancel{\text{ft}^2} \cdot 0.3 \cancel{\text{h}} \cdot (450°\cancel{\text{F}} - 72°\cancel{\text{F}})}{0.5 \cancel{\text{ft}}} \qquad \leftarrow \quad \text{all units cancel but Btu}$$

$$Q = 44 \text{ Btu}$$

Table 11.3

Thermal Conductivities

Substance	J/(s m °C)	Btu/(ft °F h)
Air	0.025	0.015
Aluminum	230	140
Brass	120	68
Brick/concrete	0.84	0.48
Cellulose fiber (loose fill)	0.039	0.023
Copper	380	220
Corkboard	0.042	0.024
Glass	0.75	0.50
Gypsum board (sheetrock)	0.16	0.092
Mineral wool	0.045	0.026
Plaster	0.14	0.083
Polystyrene foam	0.035	0.020
Polyurethane (expanded)	0.024	0.014
Steel	45	26

Figure 11.14
Insulation that is very thick and has a small thermal conductivity has a high R-value. Insulation with higher R-values insulate better.

THERMAL INSULATION

The thermal insulation you buy from a store has a value attached to it, which determines its insulating properties. This is called the *R-value*. A high R-value means it's a good insulator (Fig. 11.14).

$$R = \frac{l}{k}$$

$$R\text{-value} = \frac{\text{thickness}}{\text{thermal conductivity}}$$

For multiple layers of insulation, the total R-value is just the sum of the individual R-values:

$$R_{Total} = \sum_i \frac{l_i}{k_i}$$

$$R_{Total} = \sum_i R_i$$

Therefore heat transferred by conduction can be rewritten as:

$$Q = \frac{A \cdot t \cdot \Delta T}{R_{Total}}$$

Table 11.4
R-Values For Some Common Materials

Material	R value (ft² · °F · h/Btu)
Hardwood siding (1.0 in. thick)	0.91
Wood shingles (lapped)	0.87
Brick (4.0 in. thick)	4.00
Concrete block (filled cores)	1.93
Styrofoam (1.0 in. thick)	5.0
Fiber-glass batting (3.5 in. thick)	10.90
Fiber-glass batting (6.0 in. thick)	18.80
Fiber-glass board (1.0 in. thick)	4.35
Cellulose fiber (1.0 in. thick)	3.70
Flat glass (0.125 in. thick)	0.89
Insulating glass (0.25-in. space)	1.54
Vertical air space (3.5 in. thick)	1.01
Air film	0.17
Dry wall (0.50 in. thick)	0.45
Sheathing (0.50 in. thick)	1.32

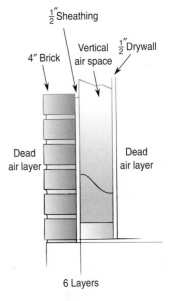

Figure 11.15

EXAMPLE 11.5

You want to insulate your home, so you make a trip to the hardware store and consider buying insulation. Comparing two types, you see an expensive one with an R-value of 30 (R-30) and a cheaper one labeled R-15. How much difference is there between the two?

R-30 is twice as big as R-15; therefore, R-30 will have half as much heat flow as R-15. In other words, R-30 is two times better at insulating than R-15.

EXAMPLE 11.6

a) Calculate the total R-value for a multilayered wall in your home if the layers consist of 4.0-in thick brick, 0.50 in of sheathing, a 3.5-in-thick vertical air space, 0.50 in of drywall, and a dead-air layer inside and outside your home. b) Calculate the heat flow through a wall measuring 10.0 ft by 8.00 ft over a 24-hour period if the average temperature is 32.0°F outside and 72.0°F inside.

a)

$$R_{brick} = 4.00 \frac{ft^2 \cdot °F \cdot h}{Btu}$$

$$R_{sheathing} = 1.32 \frac{ft^2 \cdot °F \cdot h}{Btu}$$

$$R_{air\ space} = 1.01 \frac{ft^2 \cdot °F \cdot h}{Btu}$$

$$R_{outside\ air\ film} = 0.17 \frac{ft^2 \cdot °F \cdot h}{Btu}$$

$$R_{inside\ air\ film} = 0.17\ \frac{ft^2 \cdot {}^{\circ}F \cdot h}{Btu}$$

$$R_{drywall} = 0.45\ \frac{ft^2 \cdot {}^{\circ}F \cdot h}{Btu}$$

$$R_{Total} = \sum_i R_i$$

$$R_{Total} = 7.12\ \frac{ft^2 \cdot {}^{\circ}F \cdot h}{Btu}$$

b)

$$t = 24.0\ h$$

$$T_{outside} = 32.0{}^{\circ}F$$

$$T_{inside} = 72.0{}^{\circ}F$$

$$A = 10.0\ ft \times 8.00\ ft$$

$$A = 80.0\ ft^2$$

$$Q = \frac{A \cdot t \cdot \Delta T}{R_{Total}}$$

$$Q = \frac{80.0\ \cancel{ft^2} \cdot 24.0\ \cancel{h} \cdot (72.0{}^{\circ}\cancel{F} - 32.0{}^{\circ}\cancel{F})}{7.12\ \dfrac{\cancel{ft^2} \cdot {}^{\circ}\cancel{F} \cdot \cancel{h}}{Btu}}$$

$$Q = 10.8 \times 10^3\ Btu \quad \leftarrow \quad \text{Heat flow through one wall}$$

For the house to remain at the same temperature, any heat loss must be replaced by heat generated from the furnace.

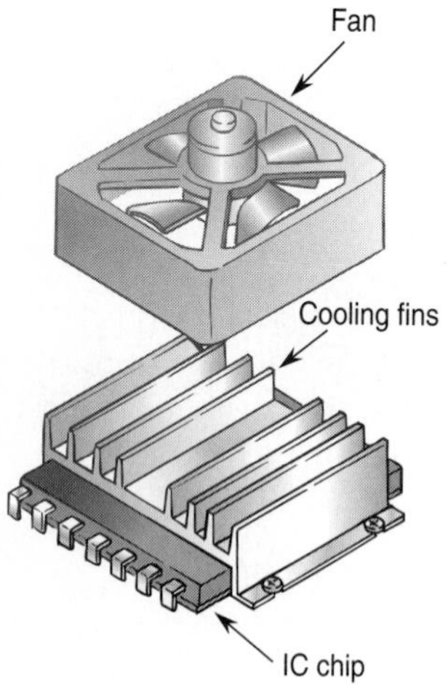

CONVECTION

Convection is heat transferred by the motion of a fluid or gas (Fig. 11.16). A gas or liquid can carry heat to or from an object.

Automobile Engine Cooling

One example of convection is the cooling system in a car (Fig. 11.17). Water circulates through the engine block, carrying heat away to the radiator. At the radiator, convection takes place with air blowing through the radiator fins, removing the heat to the atmosphere.

Home Heating

Many homes are heated with convection heating (Fig. 11.18). A fan blows air over a furnace core, transferring its heat throughout the home using heating ducts. A cold air return is provided to allow for circulation.

Solar Water Heater

A passive solar water heater does not use a mechanical pump to circulate the water (Fig. 11.19). Because hot water rises, it will automatically circulate.

Figure 11.16
Convection transfers heat through a gas or liquid. In this sketch an IC chip is cooled by convection.

Fan

Cooling fins

IC chip

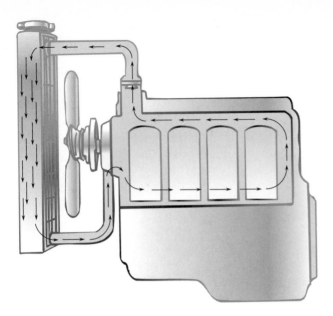

Figure 11.17
Convection cooling in an automobile.

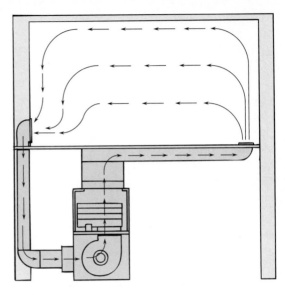

Figure 11.18
Convection heating in a home.

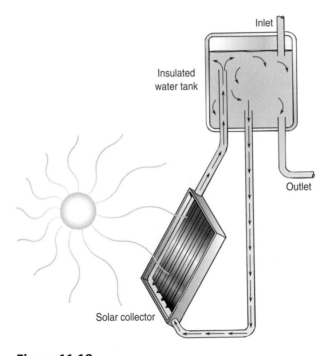

Figure 11.19
A solar water heater.

Convection depends on a number of factors:

- The faster the fluid or air flow, the quicker the heat will be transferred.
- Increasing the surface area will increase the rate of transfer; that is, big cooling fins mean faster cooling.
- Fluids or gases with high thermal conductivities transfer heat better.
- The bigger the temperature difference between the fluid or gas and the object transferring the heat, the faster the heat will get transferred; that is, cold water will cool an object faster than warm water.

Figure 11.20
Hot objects transfer heat by giving off electromagnetic radiation; cool objects warm up by absorbing electromagnetic radiation.

 RADIATION

Objects can heat up or cool down by either absorbing or emitting electromagnetic radiation (Fig. 11.20).

The sun emits light at all wavelengths. It emits radio waves, infrared, visible and even x-rays. It needs nothing to travel on or in. Like the sun, any object that is hotter than its environment emits some infrared radiation. Radiation is one of the ways a hot object cools down or a cool object warms up (Figs. 11.21–11.23).

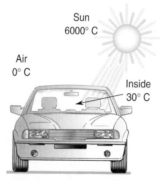

Figure 11.21
Having lots of south-facing windows in a house can reduce a heating bill because the windows allow in radiation from the sun.

Figure 11.22
Even on a cold winter's day, the inside of a car can become very warm from the sun's radiation.

Figure 11.23
A microwave oven heats food using electromagnetic radiation.

 CONTROLLING TEMPERATURE

Controlling the temperature of something like an oven requires knowing how much heat to add. Adding too much or too little heat will result in the oven being too hot or too cold. How much heat to add to reach a certain temperature will depend on how big the oven is and what it is made of. Not all materials change temperature by the same amount for the same heat input and weight. For example:

- 1 Btu of heat added to 1 Lb of steel raises the temperature 8.7°F.
- 1 Btu of heat added to 1 Lb of water raises the temperature only 1°F.

How much a material changes temperature for a given weight and heat input is determined by its specific heat capacity.

SPECIFIC HEAT CAPACITY

Specific heat capacity is the amount of heat needed to raise the temperature of a given amount of a substance by 1°. The specific heat capacity is given by the letter *c*.

It takes 1 Btu of heat added to water to raise the temperature of 1 Lb of water by 1°F.

$$C_{water} = 1\frac{Btu}{Lb°F}$$

Mathematically, the change in temperature of a material of a given weight w or mass m, specific heat capacity c and heat exchanged Q is given by:

$$\Delta T = \frac{Q}{c \cdot m} \quad \text{or} \quad \Delta T = \frac{Q}{c \cdot w}$$

$$\text{Temperature change} = \frac{\text{heat}}{\text{specific heat capacity} \cdot (\text{mass or weight})}$$

Which equation you choose depends on the units of the heat capacity. This equation says how much a material's temperature changes depends on the amount of heat exchanged and is inversely proportional to its specific heat capacity and its weight or mass. If the specific heat and weight (or mass) are very large, a lot of heat is needed to change the temperature.

EXAMPLE 11.7

How much heat is needed to raise the temperature of 198 Lb of water in a bathtub from 72.0°F to 110.0°F? How much fuel oil would be needed to produce this temperature change if fuel oil generates 17,500 Btu/Lb of heat when burned?

$$T_i = 72.0°F$$

$$T_f = 110.0°F$$

$$w = 198 \text{ Lb}$$

$$C_{water} = 1\frac{Btu}{Lb°F}$$

$$Q = ?$$

$$Q = c \cdot w \cdot \Delta T$$

$$Q = \left(1\frac{Btu}{Lb°F}\right)(198 \text{ Lb})(110.0°F - 72.0°F)$$

$$Q = 7,520 \text{ Btu}$$

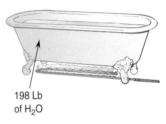

198 Lb
of H_2O

Figure 11.24

The heat of combustion of fuel oil is 17,500 Btu/Lb. To create 7,520 Btu of heat would therefore require:

$$\text{Pounds of fuel oil} = \frac{7,520 \text{ Btu}}{17,500\frac{Btu}{Lb}} \quad \leftarrow \quad \text{Note } \frac{1}{\frac{1}{Lb}} = Lb$$

Pounds of fuel oil = 0.430 Lb

EXAMPLE 11.8

A heater puts out 720 W of heat for 100.0 seconds, and all this heat goes onto 10.0 kg of steel. How much will the temperature of the steel change?

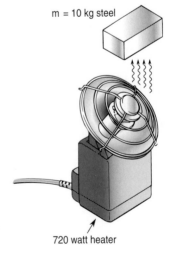

m = 10 kg steel

720 watt heater

Figure 11.25

Table. 11.5

Table of Heat Capacities

Material	Specific heat capacity	
	Btu/(Lb$_m$ °F)	J/(kg °C)
Solids		
Aluminum	0.214	895
Brass	0.094	390
Brickwork	0.20	840
Copper	0.094	390
Glass	0.194	812
Gold	0.031	130
Ice	0.504	2 110
Iron, cast	0.130	544
Lead	0.031	130
Sand	0.195	816
Silver	0.056	234
Steel	0.116	485
Stone (typical)	0.20	840
Wood, oak	0.57	2 400
Wood, pine	0.467	1 950
Liquids		
Aluminum	0.25	1 000
Copper	0.101	423
Alcohol, ethyl	0.58	2 400
Alcohol, methyl	0.60	2 500
Gold	0.032 7	137
Lead	0.038	160
Mercury	0.033	140
Nitrogen	0.474	1 980
Oxygen	0.394	1 650
Oil, machine	0.400	1 670
Silver	0.068 5	287
Water	1.000	4 184
Gases (confined)		
Air	0.168	703
Ammonia	0.399	1 670
Hydrogen	2.412	10 090
Nitrogen	0.173	724
Oxygen	0.155	649
Steam	0.346	1 450

Note: There may be some variation in these figures, depending on the actual temperature, the presence of impurities, etc.

$$P = 720 \text{ W}$$

$$t = 100.0 \text{ s}$$

$$m = 10.0 \text{ kg}$$

$$C_{steel} = 485 \ \frac{\text{J}}{\text{kg°C}}$$

$$\Delta T = ?$$

$$\Delta T = \frac{Q}{m \cdot C}$$

$$P = \frac{E}{t} \quad \rightarrow \quad E = P \cdot t$$

$$Q = E = P \cdot t$$

$$Q = (720 \text{ W})(100.0 \text{ s})$$

$$Q = 72,000 \text{ J}$$

$$\Delta T = \frac{Q}{m \cdot C}$$

$$\Delta T = \frac{72,000 \text{ J}}{(10.0 \text{ kg})\left(485\dfrac{\text{J}}{\text{kg}^\circ\text{C}}\right)}$$

$$\Delta T = 15^\circ\text{C}$$

CHANGING THE STATE OF A SUBSTANCE

Applying heat to a material doesn't always result in a temperature change. When a material is at its melting or boiling temperature, the addition of heat goes into breaking the intermolecular bonds in the material, not raising its temperature. This breaking of the intermolecular bonds will cause the material to change its *state* from a solid to a liquid (melting) or from a liquid to a gas (vaporizing). The heat needed to cause this change of state is called the *latent heat*.

Mathematically, the equation for the heat needed to change the state of a material is given by:

$$Q = m \cdot L \quad \leftarrow \quad \text{Energy needed to change state}$$
$$\text{Heat} = \text{mass} \times \text{latent heat}$$

Melting and Freezing

A solid changes to liquid when enough heat energy is delivered to it. When the melting temperature is reached, the addition of more heat does not go into raising the temperature of the solid but into breaking the intermolecular bonds in the solid. When these bonds are broken, the solid turns into liquid. This energy is called the latent heat of fusion.

Latent Heat of Fusion

Latent heat of fusion is the amount of heat needed to melt a unit amount of material when it is at its melting point (Fig. 11.26). Conversely, to freeze a liquid, energy has to be removed from it. If the liquid is at its freezing temperature, the amount of heat that must be removed from it to turn it into a solid is equal to its latent heat of fusion. The freezing temperature and melting temperature of a substance are equal.

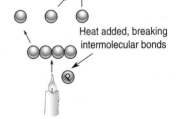

Figure 11.26
The latent heat of fusion is the energy needed to break the intermolecular bonds before a solid can change into a liquid.

Table 11.6

Latent heat of fusion and vaporization, boiling point, and freezing point of various materials

Substance	Melting Point (°C)	Latent heat of Fusion J/kg	(cal/g)	Boiling Point (°C)	Latent Heat of Vaporization J/kg	(cal/g)
Helium	−269.65	5.23×10^3	(1.25)	−268.93	2.09×10^4	(4.99)
Nitrogen	−209.97	2.55×10^4	(6.09)	−195.81	2.01×10^5	(48.0)
Oxygen	−218.79	1.38×10^4	(3.30)	−182.97	2.13×10^5	(50.9)
Ethyl alcohol	−114	1.04×10^5	(24.9)	78	8.54×10^5	(204)
Water	0.00	3.33×10^5	(79.7)	100.00	2.26×10^6	(540)
Sulfur	119	3.81×10^4	(9.10)	444.60	3.26×10^5	(77.9)
Lead	327.3	2.45×10^4	(5.85)	1750	8.70×10^5	(208)
Aluminum	660	3.97×10^5	(94.8)	2450	1.14×10^7	(2720)
Silver	960.80	8.82×10^4	(21.1)	2193	2.33×10^6	(558)
Gold	1063.00	6.44×10^4	(15.4)	2660	1.58×10^6	(377)
Copper	1083	1.34×10^5	(32.0)	1187	5.06×10^6	(1210)

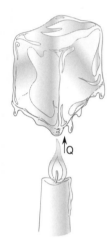

Figure 11.27

EXAMPLE 11.9

How much heat is needed to melt 2.5 kg of ice at its freezing temperature?

$$m = 2.5 \text{ kg}$$

$$L_{fusion \; for \; water} = 3.33 \times 10^5 \frac{J}{kg}$$

$$Q = m \times L$$

$$Q = 2.5 \text{ kg} \times 3.33 \times 10^5 \frac{J}{kg}$$

$$Q = 8.33 \times 10^5 \text{ J}$$

EXAMPLE 11.10

How much heat must be removed from 2.5 kg of water at its freezing temperature to turn it into ice? The amount is the same as the amount of heat needed to transform ice into water in the example above:

$$Q = 8.33 \times 10^5 \text{ J}$$

Boiling and Condensing

A liquid changes to a gas when enough heat energy is delivered to it. When the boiling temperature is reached, the addition of more heat goes not into raising the temperature of the liquid but into breaking the intermolecular bonds in the liquid. When these bonds are broken, the liquid turns into a gas. The energy needed to break these bonds is called the latent heat of vaporization.

Latent Heat of Vaporization

Latent heat of vaporization is the amount of heat needed to vaporize a unit amount of material when it is at its boiling point. Conversely, to condense a gas, in other words, turn it into a liquid, energy has to be removed from it. If the gas is at its condensing temperature, the amount of heat that must be removed to turn it into a liquid is equal to its latent heat of vaporization. The condensing temperature and boiling temperature of a substance are equal.

EXAMPLE 11.11

How much heat is needed to vaporize 1.5 kg of copper at its boiling point?

$$m = 1.5 \text{ kg}$$

$$L_{vaporization\ for\ copper} = 5.06 \times 10^6 \frac{J}{kg}$$

$$Q = m \times L$$

$$Q = 1.5 \text{ kg} \times 5.06 \times 10^6 \frac{J}{kg}$$

$$Q = 7.6 \times 10^6 \text{ J}$$

When calculating the heat needed to change a material's state when its temperature is not at its boiling or freezing temperature, the material's heat capacity must be included in the calculation.

EXAMPLE 11.12

How many joules of heat must be added to transform 12.0 kg of water from 25°C into steam?

$$m = 12.0 \text{ kg}$$

$$T_{initial} = 25.0°C$$

$$T_{boiling} = 100.0°C$$

$$Q_{Total} = Q(\text{heat needed to raise the water to } 100°C) + Q(\text{latent heat})$$

$$Q(\text{heat needed to rase the water to } 100°C) = m \cdot c \cdot \Delta T$$

$$Q = 12.0 \text{ kg} \cdot 4{,}184 \frac{J}{kg°C} \cdot (100.0°C - 25.0°C)$$

$$Q = 3.77 \times 10^6 \text{ J}$$

$$Q(\text{latent heat}) = 12.0 \text{ kg} \cdot 2.26 \times 10^5 \frac{J}{kg}$$

$$Q = 27.1 \times 10^6 \text{ J}$$

$$Q_{Total} = 3.77 \times 10^6 \text{ J} + 27.1 \times 10^6 \text{ J}$$

$$Q_{Total} = 6.48 \times 10^6 \text{ J}$$

A graph illustrating the energy needed to change 1 gram of ice to liquid, then to steam, showing the latent heats is shown in Fig. 11.28.

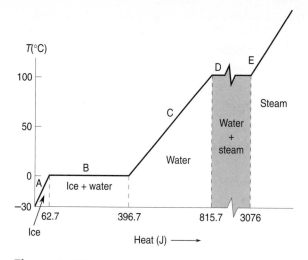

Figure 11.28
Graph of the energy needed to change 1 gram of ice to liquid and then to steam.

Pressure and the State of a Material

Pressure will also have an effect on what state a material will be in. The pressure and temperature of a material determine whether that material will be a gas, liquid, or solid. By changing these parameters, we can change the state of the material (Fig. 11.29). For example, let's look at water:

We can cause water to boil below 212°F if the pressure is low enough.

We can cause steam to liquefy above 212°F if the pressure is high enough.

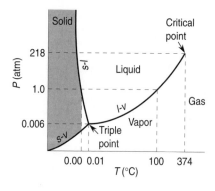

Figure 11.29
A phase diagram for water describes the state the water will be in for a given temperature and pressure.

Pressure also has an effect on evaporation and the boiling point of a liquid. Just above the surface of the water where the vapor escapes, water vapor molecules are lurking about. Under high pressure there is a better chance that an escaping water molecule will be bumped into by these molecules just above the surface. These molecules will therefore push the potential vapor back down into the liquid, reducing the evaporation or delaying the boiling point (Fig. 11.30).

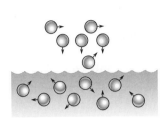

Figure 11.30
Under pressure, water molecules cannot escape easily because there is water vapor pushing them back into the liquid.

Figure 11.31
A pressure cooker delays the boiling temperature using pressure.

Pressure Cooker

An open pot of water boils at approximately 212°F. Water under pressure boils at a higher temperature. This allows food in a pressure cooker to be cooked at a higher temperature and therefore to cook faster (Fig. 11.31).

CHAPTER SUMMARY

Temperature scale conversions:

$$°C = \frac{5}{9}(°F - 32)$$

$$°F = \frac{9}{5}°C + 32$$

$$°C = °K - 273$$

$$°K = °C + 273$$

Heat is the energy transferred between two systems due to a temperature difference.
Energy conversions:

1 Btu = 1,055 J
1 Btu = 778 ft Lb
1ft Lb = 1.36 J
1 cal = 4.186 J

There are three ways to transfer heat from one place to another: conduction, convection, and radiation.
Conduction: Heat transferred between two materials in contact. Heat conduction is given by the following equation:

$$Q = \frac{k \cdot A \cdot t \cdot (T_2 - T_1)}{l}$$

Heat =
$$\frac{\text{thermal conductivity} \cdot \text{area} \cdot \text{time} \cdot (\text{temperature}_2 - \text{temperature}_1)}{\text{length}}$$

Convection: Heat transferred by the motion of a fluid or gas. A gas or liquid can carry heat to or from an object.
Radiation: Objects can either heat up or cool down by either absorbing or giving off electromagnetic radiation.
Thermal insulation: Thermal insulation is rated by the R-value. A high R-value means the material is a good insulator.

$$R = \frac{l}{k}$$

$$R\text{-value} = \frac{\text{thickness}}{\text{thermal conductivity}}$$

For multiple layers of insulation, the R-value is the sum of the individual R-values.

$$R_{Total} = \sum_i \frac{L_i}{k_i}$$

Q can be rewritten as:

$$Q = \frac{A \cdot t \cdot \Delta T}{R_{Total}}$$

Specific Heat Capacity

The amount of heat needed to raise the temperature of a given amount of a substance by 1°.

$$\Delta T = \frac{Q}{c \cdot m} \text{ or } \Delta T = \frac{Q}{c \cdot w}$$

Temperature change $= \dfrac{\text{heat}}{\text{specific heat capacity} \cdot \text{(mass or weight)}}$

The equation for the heat needed to change the state of a material is given by:

$$Q = m \cdot L$$

Heat = mass × latent heat

Problem-Solving Tip:

■ When solving problems dealing with thermal conduction, specific heat, and latent heat, look at the units for k, c, and L. They will tell you what the other units in the equation should be written in.

PROBLEMS

1. Convert 72°F to Celsius and Kelvin.

2. What happens to the kinetic energy of a gas as it is heated or cooled?

3. Suppose a heater puts out 1,200 W for 1 minute. How many Btus has it put out?

4. Suppose insulation for your attic has an R-value of 3 and a more expensive insulation has an R-value of 6. What is the difference in heat flow through these two materials?

5. Suppose you build a heated doghouse made from .5-in-thick white pine wood, which has a thermal conductivity of $\dfrac{0.065 \text{ Btu}}{\text{ft °F h}}$. The doghouse is shaped like a rectangle with dimensions of 36 in by 24 in by 28 in. You would like to maintain the temperature inside at 72°F over a period of 8 hours when the outside temperature is 32°F. What size heater do you need to install in the doghouse?

6. What if the temperature drops to 0°F in problem 5? How much more heat is needed to keep the house at 72°F?

7. How much heat is needed to heat up a tub containing 25 gal of water from 68°F to 105°F?

8. Suppose a computer puts out 210 W of heat. If this heat is not expelled to the room, the computer will continue to heat up until it is ruined. If a fan removes most of the heat by convection, how does the rest of the heat get out of the computer?

9. What gives off heat faster, a hot object or a warm object?

10. Suppose you have 2 minutes to drink a hot cup of coffee. You would like the coffee to be as cool as possible before you drink it. Should you pour the cool cream in first and wait for the coffee to cool down, or wait some time and then pour in the cool cream?

11. Suppose you build a heat source using a 100-W light bulb that puts out 20 W of light and 80 W of heat. If all this heat is put into heating 5.0 kg of water from 22°C to 75°C, how long do you have to wait for the water to reach this temperature?

12. Heat sinks mounted on electronic components dissipate heat by what method? Why are they made with fins?

13. Convert 65°F into Celsius.

14. An oven has a feedback loop that measures the temperature and compares it to the set-point temperature and turns the heating element on or off depending on whether the temperature is higher or lower than the set point. Sketch out a rough design of how you might build such a feedback circuit.

15. Soldering irons are rated in terms of wattage, not temperature. If the iron's wattage is too high,

a sensitive electronic component could be damaged by becoming overheated. Why are soldering irons rated in terms of wattage and not temperature?

16. How much heat is needed to covert 22.0 kg of water from 22.0°C into steam?

17. If the atmospheric pressure on a mountaintop is much lower than the pressure at sea level, does water on a mountaintop boil at a temperature above or below 100°C? If so, why?

Thermodynamics: Heat Engines, Heat Pumps, and Thermal Expansion

*T*hermodynamics is the physics of heat. Practial applications of thermodynamics include temperature regulation of electronics, heating and air conditioning, automobile engine design, weather prediction, and anywhere else where heat is exchanged.

Why doesn't there exist a car that gets over a hundred miles per gallon? Have the big oil companies been suppressing the patents for this technology in order to suppress the building of such efficient engines? Lets look at the physics behind these ideas.

Figure 12.1

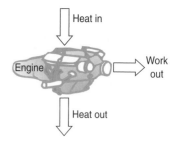

Figure 12.2
A heat engine.

 HEAT ENGINES

A *heat engine* is any device that borrows heat energy from some reservoir of heat at a high temperature and converts some of it into work plus some wasted heat at a lower temperature (Fig. 12.2).

Examples of heat engines are gas, diesel, jet, steam, and rocket engines (Fig. 12.3). They all take heat energy from fuel and convert it into some type of work to move a vehicle.

Steam engine

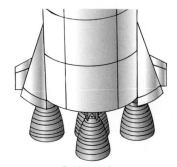
Jet engine

Fuels

Fuels store chemical potential energy. A molecule of gasoline can be thought of as a bunch of atoms separated by compressed springs. The springs are the interatomic force between atoms. With the addition of heat, the springs will be released from compression, hurling the atoms outward. This is an explosion of the gas molecule, turning the chemical potential into kinetic energy (Fig. 12.4). Table 12.1 lists the amount of energy released when these fuels are burned.

Table 12.1
Btus of Heat Produced by the Combustion Per Amount of Fuel

Fuel	Heat of combustion		Air-fuel ratio for complete combustion, Lb_m/Lb_m or g/g
	Btu/Lb_m	MJ/kg	
Butane	20 000	46	15.4
Gasoline	19 000	44	14.8
Crude oil	18 000	42	14.2
Fuel oil	17 500	41	13.8
Coke	14 500	34	11.3
Coal, bituminous	13 500	31	10.4

Fuel	Heat of combustion				Air-fuel ratio for complete combustion, ft^3/ft^3 or L/L
	On mass basis		On volume basis		
	Btu/Lb_m	MJ/kg	Btu/ft^3	kJ/L	
Hydrogen	61 400	143	275	10.3	2.4
Methane	23 800	55.2	900	33	9.6
Propane	22 200	51.5	2400	88	13.7
Natural gas	23 600	54.8	1000	37	11.9
Coke gas	23 000	53.6	1300	50	14

*Data applies to gases at 1.0 atm pressure and 15.6°C [60°F]. Actual values vary slightly depending on geographical origin. Values do not include latent heat in any water vapor formed as a product of combustion.

Note: 1 MJ/kg = 1000 kJ/kg = 1000 J/g
    1 kJ/L = 1 J/cm^3 = 1 MJ/m^3

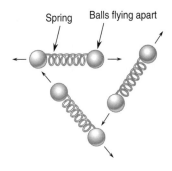
Rocket engine

Figure 12.3
Examples of heat engines.

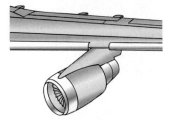

Figure 12.4
A chemical reaction releasing heat.

Gas Mileage

The gas mileage of a car is limited by how much energy an engine can extract from a gallon of gas and use it to move the car. We can increase gas mileage by making the car lighter or more aerodynamic or by making a host of other mechanical engine improvements, but there is a limit to how far we can go. A gallon of gas contains about 0.1 million Btus of heat energy. Not all of this energy is converted into moving the car. The limits are constrained by the laws of thermodynamics.

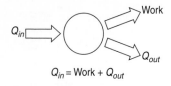

$Q_{in} = \text{Work} + Q_{out}$

Figure 12.5
The first law of
thermodynamics.

Laws of Thermodynamics

The first law of thermodynamics is about energy conservation (Fig. 12.5). You will recall that energy conservation means that the total energy of an isolated system is a constant. Energy is never created or destroyed; it just changes form. Heat is energy. Part of its input into a heat engine goes into performing some useful work, part goes into just heating up the engine, and the rest is just wasted in terms of the exhaust. A more general definition of the first law is:

First Law of thermodynamics: Energy is conserved as it moves through a thermodynamic system.

The first law applied to heat engines implies that the energy gained from a heat source equals the work done plus the heat wasted.

$$Q_{in} = W + Q_{out}$$
$$\text{Heat in} = \text{work} + \text{heat out}$$

The first law is a statement about the conservation of energy. The energy gained from some fuel source is used to do work by a heat engine, plus that wasted in just heating up the environment.

The second law of thermodynamics is related to your experience when making chocolate milk (Fig. 12.6). You start off with just milk and just chocolate syrup. When you mix the two, you get chocolate milk. Now try to separate the chocolate milk into just chocolate syrup and milk. Impossible, you say! This is the second law at work. The second law says it is easier to create a disordered mixture than to try to bring order back to the mixture. A more general way of saying this is:

Second law of thermodynamics: Systems in the universe normally evolve from order to disorder.

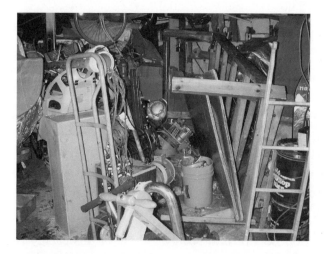

Figure 12.6
The second law of thermodynamics. The universe normally evolves from order to disorder.

When a heat engine burns fuel, it converts the internal energy of the fuel, which, when burned, produces molecules moving about randomly at high speeds into an ordered motion, moving a piston or wheel. All systems in the universe normally evolve from order to disorder. We are asking a heat engine to go against this, to turn randomness into order. The second law implies that it is impossible to build a heat engine that converts all the energy in the fuel into work. Some of the energy will be converted into the orderly motion of a piston, and the rest will go on to just heating the environment, into more disorder. In the case of an automobile, most of the energy released from the fuel goes into heating up the cooling water and the exhaust.

Efficiency

The efficiency of an engine is a measure of how much work it can do versus how much heat energy is put in:

$$\text{Efficiency} = \frac{W}{Q_{in}} = \frac{\text{work}}{\text{heat in}}$$

The most efficient engine would be one where all of the fuel's energy is extracted out, performing some useful work. Looking at the first law

$$Q_{in} = W + Q_{out}$$

solving for W gives

$$W = Q_{in} - Q_{out}.$$

Therefore

$$\text{Efficiency} = \frac{W}{Q_{in}} = \frac{Q_{in} - Q_{out}}{Q_{in}}.$$

To maximize the efficiency, we want to maximize this difference in the numerator. Heat and temperature are related to each other, and it can be shown that efficiency depends mainly on the difference between the input and exhaust temperatures. The larger the temperature difference, the more efficient the heat engine. The *Carnot engine,* named after French scientist N.L. Sadi Carnot, who first proposed it, is ideally the most efficient heat engine possible.

The efficiency of a perfect heat engine is given by:

$$\text{Efficiency} = \left(1 - \frac{T_{out}\,(°K)}{T_{in}\,(°K)}\right) \times 100$$

This equation says that if the temperature of the heat source of the engine T_{in} is high and the wasted heat that is exhausted into an environment at temperature T_{out} is low, then the efficiency will be high. Fuel burned in an automobile engine raises the temperature inside the combustion chamber to thousands of degrees. The most efficient engine of this type is one where the combustion chamber temperature is very high and the exhaust temperature is very low.

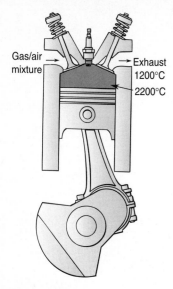

Figure 12.7

EXAMPLE 12.1

When gas is burned in a car engine, it reaches temperatures of 2,200°C and puts out exhaust at 1,200°C. What is the maximum efficiency of this engine?

$$T_{in} = 2,200°C$$

$$T_{out} = 1,200°C$$

The temperatures must be converted to kelvin.

$$°K = °C + 273$$

$$°K_{in} = 2,200°C + 273 = 2,473°K$$

$$°K_{out} = 1,200°C + 273 = 1,473°K$$

$$\text{Efficiency} = \left(1 - \frac{T_{out}}{T_{in}}\right) \times 100$$

$$\text{Efficiency} = \left(1 - \frac{1473°K}{2473°K}\right) \times 100 = 40\%$$

From this example, we can see that even a perfect heat engine that has no losses due to practical considerations such as friction will not be 100% efficient. There will still be some heat wasted to the environment (Fig. 12.8).

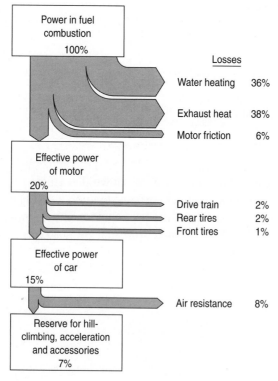

Figure 12.8
Most automobile engines have an efficiency of 25%.

Because the efficiency of any type of heat engine is always less than 100%, it follows that no one will ever build a perpetual motion machine because there will always be a need for some additional energy input without which the machine will eventually slow down and stop.

External Combustion Engine

A steam engine is an external combustion engine (Fig. 12.9). External to the engine, fuel is burned to boil the water to generate steam, which enables the engine (Fig.12.10).

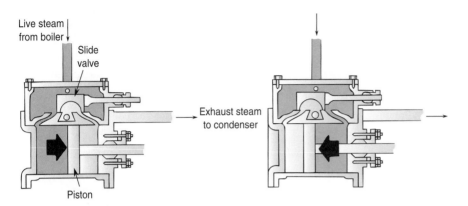

Figure 12.9
A steam engine. Valves open and close, alternately pushing the piston back and forth.

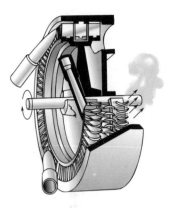

Figure 12.10
A steam turbine. The fins on a turbine are designed to extract as much thermal energy as possible from the steam and turn it into rotary motion. After exiting the turbine, the steam is cooled and returns to the boiler to become steam again and repeat the process.

Internal Combustion Engine

Fuel is burned inside the internal combustion engine. Four-stroke engines are illustrated in Figures 12.11 and 12.12.

These engines are called four-stroke because of the four-stroke cycle the engine is taken through:

1. Intake: The piston pulls the gas-air mixture into the cylinder.
2. Compression: The piston rises, compressing the fuel mixture.
3. Power: The fuel ignites, driving the piston down.
4. Exhaust: The piston pushes the burned fuel out.

The cycle then repeats itself, beginning with stroke 1.
The differences between a diesel and a gasoline engine are as follows:

- Gasoline engines use spark plugs to ignite the fuel. Diesel engines operate under much higher pressures, which ignite the fuel. Compressing any fuel enough will cause it to ignite.
- Diesel engines operate under much higher pressures, 20:1 compression ratio, compared with 8:1 for gasoline. They must be made stronger and therefore heavier than gas engines.
- Diesel engines are more efficient: about 30% efficient, compared with 20% for gas engines.

To obtain maximum efficiency/power:

- Increasing the volume of the cylinders can increase the power of an engine; this will allow for more fuel/air mixture to burn. Increasing the amount of fuel/air

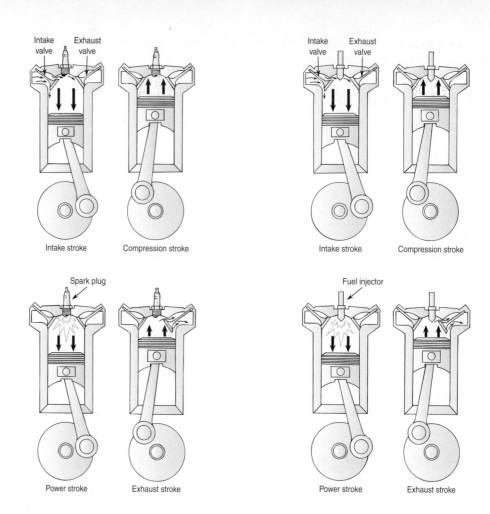

Intake valve Exhaust valve

Intake stroke Compression stroke

Spark plug

Power stroke Exhaust stroke

Figure 12.11
Four-stroke gasoline engine

Intake valve Exhaust valve

Intake stroke Compression stroke

Fuel injector

Power stroke Exhaust stroke

Figure 12.12
Four-stroke diesel engine

mixture in the cylinder can also be done by pumping the mixture into the cylinder. This is known as *supercharging*.

■ Increasing the fuel/air pressure in a cylinder by reducing the clearance between the piston and the valves can increase the power. There is a limit, however, based on the strength of the material the engine is made of.

The Efficiency of a Human Being

Food is a form of fuel. Its unit of energy is the Calorie. One food Calorie equals 1,000 physics calories (1 C = 1,000 c). One food Calorie equals 4,186 J. *The human body is about 25% efficient* meaning that only 25% of the energy we get from food is turned into muscular energy. This is about the same efficiency as that of a gasoline automobile engine. Let's analyze the energy needed over the period of one day. A man weighing approximately 155 Lb will burn 100 Cal/h just to maintain life (breathing, circulating blood, etc.) in normal waking conditions, and 60 Cal/h sleeping (Fig. 12.13).

1. 100 Cal/h for 16 waking hours equals 1,600 Cal.
2. Sleeping for 8 hours at 60 Cal/h equals 540 Cal.

Figure 12.13

3. Walking slowly, at 2.5 mph, burns 220 Cal/h. If we approximate the day's activities to walking 5 miles, that would total 440 Cal.

Total calories burned = 2,580 Cal
Therefore, just to maintain this man's weight, he must consume

4. $Q_{in} = W + Q_{out} = 2,580$ Cal every day just to maintain weight.

Compare 440 Cal burned by walking 5 miles with eating a Big Mac hamburger at 530 Cal. It is clear that if you want to lose weight, reducing your food intake will benefit you a lot more than exercise.

 ## HEAT PUMPS

Heat pumps pump heat from a low-temperature region to a high-temperature region (Fig. 12.14).

Figure 12.14
Heat pumps pump heat
from a cold to a hot region.

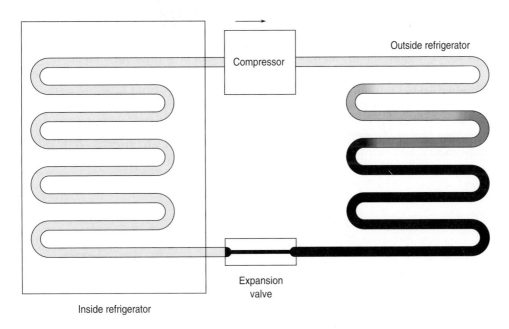

Figure 12.15
A refrigerator is a heat pump that draws heat from a low-temperature region to a higher-temperature region.

When a gas expands and evaporates, it cools. Anything coming in contact with this gas will cool down as well. This is the principle behind a refrigerator, which is an example of a heat pump (Fig. 12.15).

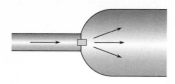

Figure 12.16
Expanding gas cools down.

Refrigeration Cycle

1. A liquid refrigerant such as ammonia or dichlorodifluoromethane with a very low boiling point (around −27°F) passes though an expansion valve like the atomizer on a spray gun. As this mist expands and evaporates, it cools (Fig. 12.16). This cool gas is run through coils in a refrigerator, cooling the inside.

2. At the other end of this coil is a compressor. The compressor draws in this gas and outputs it at a higher pressure and temperature (Fig. 12.17).

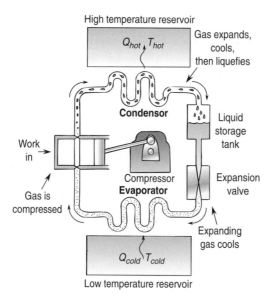

Figure 12.17
The gas is compressed, raising its temperature and pressure. The high-pressure, high-temperature gas expands, cools down, and turns back into liquid.

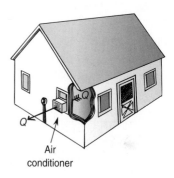

Figure 12.18
Air conditioners are basically refrigerators without the insulated box.

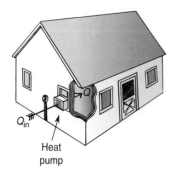

Figure 12.19
Heat pumps for heating. A heat pump can absorb heat from the outside during the winter and deliver it to the inside, where it is warmer. It is basically an air conditioner turned around.

3. This high-pressure, high-temperature gas passes through a set of coils on the back of the refrigerator. Passing through the coils, it emits this heat to the room and therefore lowers its temperature. As the gas cools, it turns back into a liquid.

4. The cycle repeats itself.

Heat pumps can be used to cool a house, as in the case of an air conditioner (Fig. 12.18), or warm a house (see Fig. 12.19).

A *Peltier device* is an electronic refrigerator made from several metal-semiconductor junctions in series with an alternating array of n-type and p-type semiconductors (Fig. 12.20). Current flowing through this device results in one side getting hot, the other cold. These devices can be found in electronic coolers.

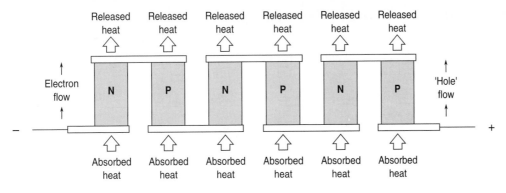

Figure 12.20
A Peltier device is an electronic refrigerator made from several metal-semiconductor junctions in series with an alternating array of n-type and p-type semiconductors.

THERMAL EXPANSION OF GASES, LIQUIDS, AND SOLIDS

Most materials expand when heated. Thermal expansion properties of materials can be exploited to build temperature switches, valves, refrigerators, air conditioners, and a host of other useful devices. Knowledge of how much they expand is important in designing anything that cycles through temperature ranges such as bridges, roads, or even printed circuit boards (Fig. 12.21).

Figure 12.21
An expansion joint. As the seasons change, the bridge will expand in the heat and shrink in the cold.

Linear Expansion

A change in temperature causes most materials to change length (Fig. 12.22). This can be expressed mathematically as:

Figure 12.22
Area and volume thermal expansion.

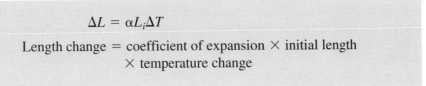

$$\Delta L = \alpha L_i \Delta T$$

Length change = coefficient of expansion × initial length × temperature change

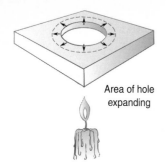

Area of hole
expanding

Area and Volume Expansion

The thermal expansion of an area of a material is twice as big as the linear expansion, and the volume expansion is three times as big (Fig. 12.23).

$$\Delta A = 2\alpha A_i \Delta T \quad \leftarrow \quad \text{area expansion}$$

$$\Delta V = 3\alpha V_i \Delta T \quad \leftarrow \quad \text{volume expansion}$$

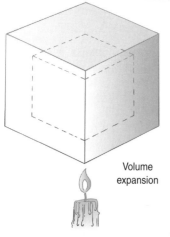

Volume
expansion

Figure 12.23
Linear thermal expansion.

Table 12.2
Coefficients of Linear Expansion

Material	α (metric)	α (English)
Aluminum	$2.3 \times 10^{-5}/\text{C}^\circ$	$1.3 \times 10^{-5}/\text{F}^\circ$
Brass	$1.9 \times 10^{-5}/\text{C}^\circ$	$1.0 \times 10^{-5}/\text{F}^\circ$
Concrete	$1.1 \times 10^{-5}/\text{C}^\circ$	$6.0 \times 10^{-6}/\text{F}^\circ$
Copper	$1.7 \times 10^{-5}/\text{C}^\circ$	$9.5 \times 10^{-6}/\text{F}^\circ$
Glass	$9.0 \times 10^{-6}/\text{C}^\circ$	$5.1 \times 10^{-6}/\text{F}^\circ$
Pyrex	$3.0 \times 10^{-6}/\text{C}^\circ$	$1.7 \times 10^{-6}/\text{F}^\circ$
Steel	$1.3 \times 10^{-5}/\text{C}^\circ$	$6.5 \times 10^{-6}/\text{F}^\circ$
Zinc	$2.6 \times 10^{-5}/\text{C}^\circ$	$1.5 \times 10^{-5}/\text{F}^\circ$

EXAMPLE 12.2

A steel tape measure is 100 ft at 72°F. The coefficient of linear expansion for steel is $6.5 \times 10^{-6}/°F$. How much did the tape measure change length at 5°F?

$$L_i = 100 \text{ ft}$$

$$\alpha = 6.5 \times 10^{-6}/°F$$

$$\Delta T = 72°F - 5°F = 67°F$$

$$\Delta L = \alpha L_i \Delta T$$

$$\Delta L = (6.5 \times 10^{-6}/°\cancel{F}) \cdot 100 \text{ ft} \cdot 67°\cancel{F}$$

$$\Delta L = 0.04 \text{ ft}$$

$$\Delta L = 0.5 \text{ in} \quad \leftarrow \quad \text{The length will change by as much as half an inch!}$$

Figure 12.24

EXAMPLE 12.3

A brass bushing with an inside diameter of 1.980 in at 72°F is to be sweat fit onto a shaft of diameter 2.000 in. Sweat fitting is a process of increasing the size of something by heating it. What temperature must the bushing be brought to before it will fit?

$$\text{Initial diameter} = 1.980 \text{ in}$$

$$\text{Final diameter} = 2.000 \text{ in}$$

$$\alpha_{\text{brass}} = 1.0 \times 10^{-5}/°F$$

$$\Delta L = \alpha L_i \Delta T$$

$$\Delta L = 2.00 \text{ in} - 1.980 \text{ in} = 0.020 \text{ in}$$

$$\Delta L = \alpha L_i \Delta T$$

$$\frac{\Delta L}{\alpha L_i} = \frac{\cancel{\alpha} L_i \Delta T}{\cancel{\alpha} \cancel{L_i}} \quad \leftarrow \quad \text{solving for } \Delta T$$

$$\Delta T = \frac{\Delta L}{\alpha L_i}$$

$$\Delta T = \frac{0.020 \text{ in}}{(1.0 \times 10^{-5}/°F)(1.98 \text{ in})}$$

$$\Delta T = 1,010°F$$

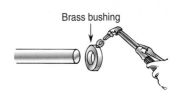

Figure 12.25

Bimetal Strip Thermostat

A *bimetal strip thermostat* is essentially a heat switch consisting of two types of metal bonded together, with electrical connections on either end. As the temperature changes, the two metals expand at different rates, causing the strip to bend and thereby closing the switch (Fig. 12.26). Varying the spring tension that is holding the switch closed can set the operating temperature of the switch.

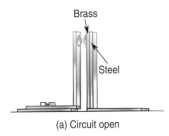

(a) Circuit open

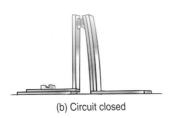

(b) Circuit closed

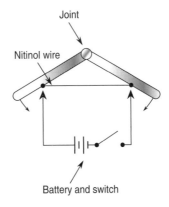

Figure 12.26
A thermal switch consists of two dissimilar pieces of metal that have different coefficients of thermal expansion. When heated, the result is a bending of the strip.

Figure 12.27
Nitinol wire contracts when heated, lending itself to actuating arms or levers. Heating takes place when enough current flows through the wire.

Nitinol Wire

Nitinol wire is an alloy of nickel and titanium that is made specifically for actuating mechanical arms. It has a very large negative coefficient of expansion, that is, it contracts when heated. When heated, these wires contract as much as 20%. Running enough current through them will cause them to heat and therefore contract (see Fig. 12.27 on the previous page). Turning off the current allows the wire to cool and slowly return to its original length. These wires can be used like motors to move arms or levers.

CHAPTER SUMMARY

First law of thermodynamics: Energy is conserved as it moves through a thermodynamic system.

The first law applied to heat engines implies that the energy gained from a heat source equals the work done plus the heat wasted.

$$Q_{in} = W + Q_{out}$$

Heat in = work + heat out

Second law of thermodynamics: All systems in the universe normally evolve from order to disorder.

The efficiency of an engine is a measure of how much work it can do compared with how much heat energy is put in.

$$\text{Efficiency} = \frac{W}{Q_{in}} = \frac{\text{work}}{\text{heat in}}$$

Ideal efficiency: The efficiency of a perfect heat engine is given by

$$\text{Efficiency} = \left(1 - \frac{T_{out}(°\text{K})}{T_{in}(°\text{K})} \right) \times 100$$

Linear expansion: A change in temperature causes most materials to change length. This can be expressed mathematically as:

$$\Delta L = \alpha L_i \Delta T$$

Length change = coefficient of expansion × initial length × temperature change

Area and volume expansion: The thermal expansion of an area of a material is twice as big as the linear expansion, and the volume expansion is three times as big.

$$\Delta A = 2\alpha A_i \Delta T \quad \leftarrow \quad \text{area expansion}$$
$$\Delta V = 3\alpha V_i \Delta T \quad \leftarrow \quad \text{volume expansion}$$

Heat pumps: Heat pumps pump heat from a low-temperature region to a high-temperature region.

Problem-Solving Tips

- When calculating ideal efficiency, remember to calculate in degrees kelvin.

- When solving problems dealing with thermal expansion, look at the units of the coefficient of thermal expansion when determining what units the other variables in the equation should be in.

PROBLEMS

1. If an engine performs 125,000 J of work using a fuel containing 620,000 J of energy, how much energy is wasted? What is the efficiency?

2. Why can't a perpetual motion machine ever be invented?

3. Compare two engines, one with an exhaust temperature of 1,600°C and the other with an exhaust temperature of 1,100°C. Which engine is more efficient?

4. Determine your calorie intake for one day and approximate the amount of work you perform. What is your efficiency?

5. A piece of brass 12.00 in long is heated and changes temperature by 123°C. What is its change in length?

6. A piece of steel has a hole in it with a diameter of 1.00 in. A pipe is to fit into it with a diameter of 1.05 in. How high must the steel be heated so that the pipe will fit into the hole?

7. What is the purpose of an expansion joint on a bridge?

8. How does a thermostat work?

9. A thermal circuit breaker uses a bimetal strip to open a circuit in the event of an overload. How does this work?

10. How does an air conditioner work? Provide a diagram.

11. Give an example of the second law of thermodynamics from everyday life.

12. Suppose you connect an electric motor to a generator and use the output from the generator to run the motor. Will this work? What do the laws of thermodynamics say about this?

13. If a heat engine takes in 2,900 J of heat and expels 1,800 J, how much work did it perform? What is its efficiency?

14. What is the difference between an internal and an external combustion engine?

15. Suppose a Peltier device has a cooling power of 40 W. How long will it take to cool 1.0 kg of water from 30°C to 10°C?

16. How could you use an air conditioner to heat your house in the winter?

Waves

This chapter will describe waves in general, but will focus on sound waves in particular. Common types of waves are water, light, sound, and radio waves (Fig. 13.1). Although they are very different from one another, they share many characteristics. With the whole world going wireless, knowledge of waves and their properties is very important.

Figure 13.1
Water waves.

Figure 13.2
The creation of a water wave.

To create a wave, you must disturb the medium. In the case of water, when you tap it with a stick, the disturbance will propagate outward (Fig. 13.2).

The actual water molecules touched by the stick do not travel outward. Disturbing the water merely pushes on the water adjacent to it, and then that water pushes on the water next to it, and so on. It is like dominoes falling over and causing the others to fall over. There is motion, but there is no transport of matter (Fig. 13.3). It is just the disturbance that moves. The wave carries energy but no matter.

Figure 13.3
Dominoes behave in a wavelike manner, propagating a disturbance. No one domino gets transported; only the energy is transported.

VELOCITY, FREQUENCY, AND WAVELENGTH

Creating waves on water requires disturbing the water. For example, tapping the water with a stick at some rate or frequency will produce a wave with the same frequency. The frequency of the waves is related to the period, or time, between waves. This relationship is given by:

$$\text{Frequency} = \frac{1}{\text{period}}$$

Frequency has units of hertz (Hz) or cycles per second.

$$1 \text{ Hz} = 1\frac{\text{cycle}}{\text{s}} = \frac{1}{\text{s}}$$

A wave has a length called a wavelength (λ), a rate or frequency (f) at which it oscillates, and a speed (v) at which it travels (Fig. 13.4). All waves, including light, sound, and radio, obey the following equation:

$$v = \lambda \cdot f \quad \leftarrow \quad \text{For all types of waves}$$

$$\text{velocity} = \text{wavelength} \times \text{frequency}$$

Figure 13.4
A wave of wavelength λ, a period T, and a speed v.

EXAMPLE 13.1

The musical note A has a frequency of 440 Hz. If the sound wave is traveling at 343 m/s, what is its wavelength?

$$f = 440 \text{ Hz}$$

$$v = 343 \text{ m/s}$$

$$\lambda = ?$$

$$v = \lambda \cdot f$$

$$\lambda = \frac{v}{f} \quad \leftarrow \quad \text{solving for } \lambda \text{ gives:}$$

$$\lambda = \frac{343\frac{\text{m}}{\text{s}}}{440 \text{ Hz}}$$

$$\lambda = \frac{343\frac{\text{m}}{\cancel{s}}}{440\frac{1}{\cancel{s}}}$$

$$\lambda = 0.78 \text{ m}$$

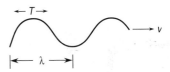

$f = 440 \text{ Hz}$
$\lambda = ?$

343 m/s

Figure 13.5

Figure 13.6
A jump rope demonstrating transverse waves

There are two types of waves, transverse and longitudinal.

Transverse Waves

Transverse waves oscillate perpendicular to the direction of travel of the wave (Fig. 13.6). Light is a transverse wave (Fig. 13.7).

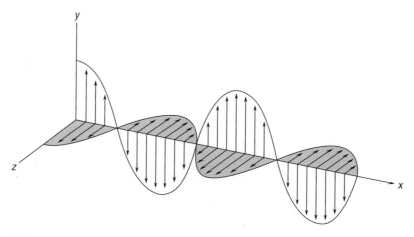

Figure 13.7
Light is a transverse wave. It is made up of electric and magnetic fields that oscillate perpendicular to the direction of travel.

Longitudinal Waves

Longitudinal waves oscillate back and forth in the direction the wave is traveling (Fig. 13.8). Sound is a longitudinal wave.

Some waves can be both transverse and longitudinal, as in the case of water (see Fig. 13.9).

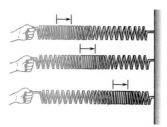

Figure 13.8
Longitudinal waves oscillate back and forth in the direction the wave is traveling.

Figure 13.9
Water is both a longitudinal and a transverse wave. A cork will bob up and down (the transverse component) and will also move back and forth (the longitudinal component).

Refraction

Refraction describes the bending of a wave when it moves from one medium into another. Light and sound waves refract because the speed of the wave changes when the wave moves from one medium into another. For example, light travels slower through water than through air. A wave front, striking the surface of water from air, will bend because the speed of the wave changes when the wave enters the water. The wave front in the water will lag behind the incoming wave. As a result, the wave will bend (Fig. 13.10).

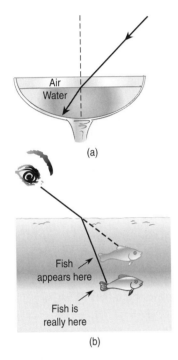

(a)

(b)

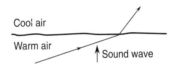

Figure 13.10
Light moving from air to water. a) Light bends because it slows down in the water and cannot keep up with the incoming wave. b) Objects appear shallower in the water because of refraction.

Figure 13.11
Sound waves bend when entering a different temperature region.

Sound waves exhibit refraction when moving from warm air of relatively low density to a cooler, denser air (Fig. 13.11). The speed of sound in warm air is faster than in cool air. As a result, the wave that enters the cool air falls behind the wave in the warm air. The sound wave is effectively bent because of this slowing down.

Superposition

Many waves can travel through the same medium without affecting each other. The overall wave will be the sum of all the waves. This is called *superposition*. Light, sound, and water waves all obey superposition. For example, in the past, there were dedicated phone lines where many conversations took place on the same wire at the same time (Fig. 13.12). The individual conversations were modulated at different frequencies. If you looked at the waveform coming out of the phone line with many conversations at once, it would look like one big messy wave.

Figure 13.12
A superposition of many signals on one wire.

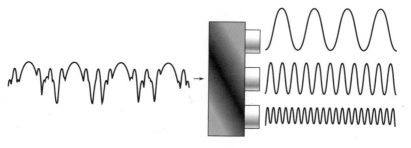

Figure 13.13
A filter separating out the individual waves.

The individual waves, or conversations, can be separated because they operate at different frequencies. A filter can be used to separate them (Fig. 13.13).

Constructive Interference

Waves that are in sync with one another combine to make a bigger wave. This is known as *constructive interference* (Fig. 13.14).

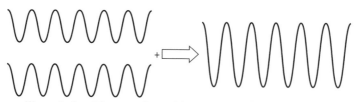

Waves starting at the same place, adding together to make a bigger wave.

Figure 13.14
Constructive interference.

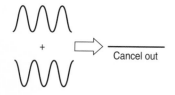

Figure 13.15
Destructive interference.

Destructive Interference

Waves that are out of sync with one another subtract to make a smaller wave. This is known as *destructive interference* (Fig. 13.15).

A *spectrum analyzer* is used to separate the individual waves that make up a waveform (Fig. 13.16).

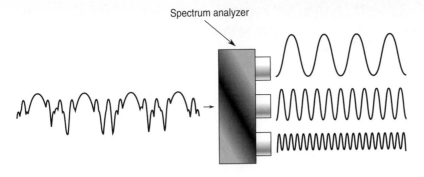

Spectrum analyzer

Figure 13.16
A spectrum analyzer determines the individual waves that make up a complex waveform.

Tone

Multiple waveforms, acting together, make up the *tone* of a sound (Fig. 13.17). The waves making up one sound may have different frequencies, phases, and intensities.

Figure 13.17
Most sound waves are a composite of many different waves.

Resonance

Resonance is maximum energy transfer (Fig. 13.18). For example, a singer can shatter a glass if the volume and frequency of his or her voice are just right. Everything has a natural frequency where, if vibrated, it will begin to shake a lot. This frequency is called the *resonance frequency*. Antennas emit the most power at their designed resonance frequency.

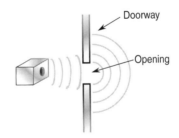

Doorway

Opening

Figure 13.18
Example of resonance: maximum energy transfer

Figure 13.19
An example of diffraction is sounds traveling through openings such as doorways.

Diffraction

Diffraction is the bending of waves around obstacles. Diffraction occurs most strongly when the size of the obstacle is the size of the wavelength of the wave (Fig. 13.19).

Figure 13.20
Radar.

DISTANCE AND VELOCITY MEASUREMENTS WITH WAVES

Radar and Sonar

Radar and sonar use radio and sound waves, respectively, to determine the distance to an object (Figs. 13.20 and 13.21). When a wave is emitted, a clock is started and times how long it takes for the wave to travel, reflect off an object, and return. Radar is typically used for long-distance determinations such as by air traffic controllers to locate airplanes. Sonar moves much more slowly than radar, which is why it is better used for short-distance applications. Underwater, radar fails because radio waves are heavily absorbed in water. Sonar is therefore used by submarines and fisherman.

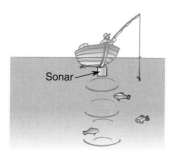

Figure 13.21
Sonar.

Doppler Shift

If a source of sound or light is approaching you, the frequency of the waves will increase; if the source is moving away from you, the frequency will decrease (Fig. 13.22). Consider an approaching train. You hear the pitch of the engine increasing as the train approaches, but as the train passes, the pitch lowers. This is because the sound waves become compressed as the source of sound approaches you and expand when the source moves away from you. This change in pitch or frequency due to a relative motion of the source is known as a *Doppler shift.* How much the pitch changes depends on the speed of the source.

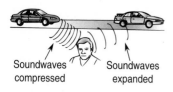

Figure 13.22
The Doppler shift of approaching and receding sources.

An application of the Doppler effect is a police *radar gun,* which consists of a microwave radio source emitting a beam of radio waves at a given frequency (Fig. 13.23). The frequency of radio waves increases after the waves bounce off oncoming objects. This is because the wave becomes compressed. The reflected wave returns to the gun, and its frequency is compared with the frequency of the original wave. The difference in the frequencies of the two waves is a measure of how fast the car is moving. Doppler radar is used to determine the speed of clouds (Fig. 13.24).

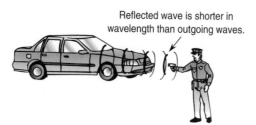

Reflected wave is shorter in wavelength than outgoing waves.

Figure 13.23
Police radar using the Doppler effect to determine speed.

Figure 13.24
Weather forecasters use Doppler radar to determine the speed of clouds.

The Doppler effect can be applied to ultrasonic waves to determine speed (Fig. 13.25). Speed can be determined by bouncing an ultrasonic sound wave off of a moving object, then measuring the change in frequency between the outgoing and incoming waves.

Mathematically, the frequency shift from the Doppler effect for light and sound is given by:

Ultrasonic doppler gun

For light,

$$f_o = f_s \cdot \frac{\sqrt{1 - v_s^2/c^2}}{1 \pm v_s/c} \quad \leftarrow \quad + \text{ sign for receding sources,}$$

$$- \text{ sign for approaching sources:}$$

For sound,

$$f_o = f_s \cdot \left(\frac{v \pm v_o}{v \mp v_s} \right) \quad \leftarrow \quad \frac{-}{+} \text{ sign for a source or observer moving away from each other}$$

$$\frac{+}{-} \text{ sign for a source or observer moving toward each other}$$

Where

f_o = frequency measured by the observer, f_s = frequency of the source

c = the speed of light, v = the speed of sound

v_s = speed of the source, v_0 = speed of the observer

Figure 13.25
The Doppler effect can be applied to ultrasonic waves to determine speed.

EXAMPLE 13.2

A train is approaching you at 22.0 m/s and is blowing a whistle at a frequency of $44\overline{0}$ Hz. Suppose the speed of sound is 345 m/s. What does the frequency of the whistle sound like to you?

$f_s = 440$ Hz

$v = 345$ m/s

$v_s = 22.0$ m/s

$v_o = 0$

$f_o = ?$

$$f_o = f_s \cdot \left(\frac{v}{v - v_s} \right) \quad \leftarrow \quad \text{Source is approaching, so we choose the } - \text{ sign.}$$

$$f_o = 44\overline{0} \text{ Hz} \cdot \left(\frac{345 \text{ m/s}}{345 \text{ m/s} - 22.0 \text{ m/s}} \right) = 47\overline{0} \text{ Hz}$$

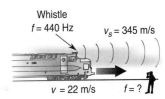

Whistle
$f = 440$ Hz $v_s = 345$ m/s

$v = 22$ m/s $f = ?$

Figure 13.26

EXAMPLE 13.3

Some police radar guns use a microwave transmitter that sends out frequencies in the X-band of 10.525 GHz or the K-band centered at 24.150 GHz. Suppose a police officer is shooting radar with a radar gun that sends out a frequency of 24.150 GHz. If a car is approaching at 100 mi/h, what does the return frequency measure?

$v = 100$ m/hr

Figure 13.27

$$f_s = 24.150 \text{ GHz}$$

$$c = 186{,}000 \text{ mi/s}$$

$$v_s = 100.0 \text{ mi/h}$$

$$f_o = ?$$

$$1 \text{ h} = 3{,}600 \text{ s}$$

$$186{,}000 \, \frac{\text{mi}}{\cancel{s}} \left(\frac{3{,}600 \, \cancel{s}}{1 \text{ h}} \right) = 6.70 \times 10^8 \text{ mi/h}$$

$$f_o = f_s \cdot \frac{\sqrt{1 - v_s^2/c^2}}{1 - v_s/c} \qquad \leftarrow \quad - \text{ sign for approaching}$$

$$f_o = 24.150 \text{ GHz} \cdot \left(\frac{\sqrt{1 - \dfrac{(100.0 \text{ mi/h})^2}{(6.70 \times 10^8 \text{ mi/h})^2}}}{1 - \dfrac{100.0 \text{ mi/h}}{6.70 \times 10^8 \text{ mi/h}}} \right)$$

$$f_o = 24.1500036 \text{ GHz}$$

Note: We have neglected the rules of significant digits in this example.

Figure 13.28
A speaker cone moves back and forth, creating sound. Sound is the compression and decompression of air traveling out from the source.

SOUND WAVES

Sound is a longitudinal wave. Sound waves need a medium through which to travel, such as air, solids, or liquids. To create a sound wave, we need to somehow disturb the medium. For example, a stereo speaker creates sound by moving its cone back and forth (Fig. 13.28). As the speaker cone moves forward, it compresses the air; when it moves backward, it creates a slight vacuum. The speaker is alternately compressing and decompressing the air. The volume of air disturbed by the speaker does not travel outward. It merely pushes on the air next to it, and then that air pushes on the air next to it. It is just the disturbance that moves.

Speed of Sound

Sound typically travels fastest through solids, then liquids, and slowest through gases.

Table 13.1
The Speed of Sound Through Media

Medium	Speed	
	m/s	ft/s
Aluminum	6,420	21,100
Brass	4,700	15,400
Steel	5,960	19,500
Granite	6,000	19,700
Alcohol	1,210	3,970
Water (25°C)	1,500	4,920
Air, dry (0°C)	331	1,090
Vacuum	0	0

The speed of sound changes with temperature, pressure, and humidity of the air. At 72°F in dry air, at 1 atmosphere (atm) of pressure, the velocity is:

$$v_s = 345\,\frac{m}{s} \text{ or } v_s = 1{,}132\,\frac{ft}{s} \text{ or } v_s = 772\,\frac{mil}{h}$$

In general the speed of sound changes with temperature according to:

$$v = 331\frac{m}{s} + \left(0.6\,\frac{m/s}{°C}\right) \cdot T_c$$

Where v is in m/s and T_c is in Celsius, the speed of sound increases with temperature.

EXAMPLE 13.4

What is the velocity of sound at 31°C?

$$T_c = 31°C$$

$$v = 331\frac{m}{s} + \left(0.6\,\frac{m/s}{°C}\right) \cdot T_c$$

$$v = 331\frac{m}{s} + \left(0.6\,\frac{m/s}{\cancel{°C}}\right) \cdot 31\cancel{°C}$$

$$v = 350 \text{ m/s}$$

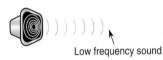

High frequency sound

Low frequency sound

Figure 13.29
A speaker generating high- and low-frequency sounds.

Pitch

The *pitch* of sound is the same as the frequency. In terms of generating sound with a speaker, the pitch is determined by how fast the speaker cone is moving back and forth. If the cone is oscillating back and forth at a high frequency, the pitch will be high (Fig. 13.29).

The human ear can hear sounds that range from a frequency of 20 Hz to 20 kHz. Sounds below this range are called *infrasonic*. Sounds above this range are called *ultrasonic*.

Low-frequency sounds spread out from a source, that is, they are not very directional (Fig. 13.30). They are not readily absorbed, and as a result, they travel through walls easily. This is familiar to anyone living next door to someone with a stereo.

High-frequency sounds are more directional than low-frequency sounds (Fig. 13.31). Because of this, in the ultrasonic region, there are many applications where the waves can be focused.

Figure 13.30
Low-frequency sounds spread out from a source.

Ultrasound Imaging

Obstetricians use *ultrasound imaging* to see a fetus while it is still inside the mother (Fig. 13.32). Sound waves, with frequencies of 3.5 to 7 MHz, are sent into

Figure 13.31
High-frequency sounds are directional.

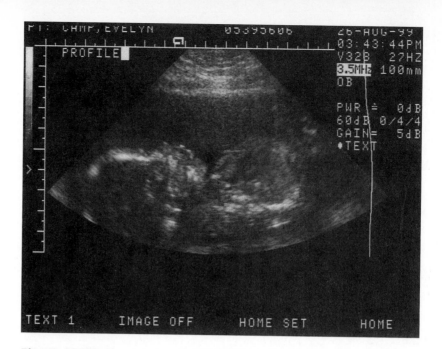

Figure 13.32
Ultrasound imaging.

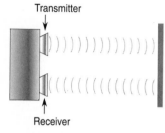

Figure 13.33
The ultrasonic tape measure works by sending a sound wave and timing how long it takes to return after bouncing off an object whose distance is being measured.

the mother's abdomen. Sound travels at about 1,540 m/s in soft tissue. Quartz crystals generate the sound and also act as the receiver of the returning signals. These waves are reflected and absorbed by the soft tissue. The reflected waves returning to the ultrasound transducer are then put together to form the image of the fetus. The higher-frequency waves have higher resolution than the low-frequency waves, but they do not penetrate as deeply.

Ultrasonic Tape Measure

An *ultrasonic tape measure* measures the distance to objects by bouncing an ultrasonic wave off the object and timing how long it takes for the wave to come back. The longer it takes, the farther away the object (Fig. 13.33).

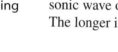

Figure 13.34
Ultrasonic cleaning.

Ultrasonic Cleaning

Ultrasonic cleaning uses sound waves above 20 kHz to vibrate dirt loose from objects being cleaned (Fig. 13.34). The sound causes the object to vibrate at the same frequency, shaking the dirt and grime loose.

Sound Pressure

A sound wave is essentially a pressure wave, that is, a compression and decompression of air that travels. When this pressure wave impacts the eardrum, the eardrum sends a signal to the brain, and this signal is interpreted as sound. There is energy in this pressure wave. The rate at which this energy passes through a given area is given by the intensity.

Sound Intensity

The amount of power delivered by waves per unit area is given by:

$$I = \frac{P}{A}$$

$$\text{Intensity} = \frac{\text{power}}{\text{area}}$$

See Fig. 13.35.

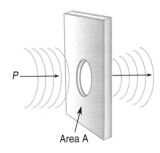

Figure 13.35
Sound of power P passing
through an area A.

If we consider a point source of sound radiating isotropically (equally in all directions) outward (Fig. 13.36), then the sound radiates outward spherically and the intensity is given by:

$$I = \frac{P}{A}$$

$$I = \frac{P}{4\pi r^2}$$

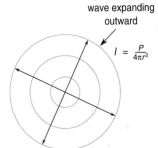

Figure 13.36
Sound radiating
isotropically (equally in all
directions) outward.

EXAMPLE 13.5

A 1.0-W point source of sound radiates outward isotropically. What is the intensity 3.0 m away from the source?

$$P = 1.0 \text{ W}$$

$$r = 3.0 \text{ m}$$

$$I = \frac{P}{4\pi r^2}$$

$$I = \frac{1.0 \text{ W}}{4\pi (3.0 \text{ m})^2}$$

$$I = 8.8 \times 10^{-3} \frac{\text{W}}{\text{m}^2}$$

Sound Intensity Level

The threshold intensity for hearing is $I_o = 1 \times 10^{-12}$ W/m². A sound that has twice the intensity as another does not sound twice as loud to the human ear. The ear's response to sound is said to be logarithmic. A sound that is twice as intense as another does not sound twice as loud but only slightly louder. The volume control on a stereo is a logarithmic potentiometer, which takes into account the logarithmic nature of our hearing.

Sound intensity level is measured in decibels (dB) and is given by:

$$dB = 10 \cdot \log\left(\frac{I}{1 \times 10^{-12}\,\text{W/m}^2}\right)$$

$$\text{Sound intensity level} = 10 \cdot \log\left(\frac{\text{intensity}}{\text{hearing intensity threshold}}\right)$$

Table 13.2
Typical Sound Levels

Jet takeoff (60 m)	120 dB	
Construction site	110 dB	Intolerable
Shout (1.5 m)	100 dB	
Heavy truck (15 m)	90 dB	Very noisy
Urban stree	80 dB	
Automobile interior	70 dB	Noisy
Normal conversation (1 m)	60 dB	
Office, classroom	50 dB	Moderate
Living room	40 dB	
Bedroom at night	30 dB	Quiet
Broadcast studio	20 dB	
Rustling leaves	10 dB	Barely audible
	0 dB	

Sound levels can be measured with a decibel meter (Fig. 13.37).

From the sound intensity level equation, we can say that a sound that is 10^N (where N is an integer) times more intense than the threshold of hearing means that $SL = N \times 10$ dB.

- A sound that is 10 times more intense than threshold means $SL = 10$ dB.
- A sound that is 100 times more intense than threshold means $SL = 20$ dB.
- A sound that is 1000 times more intense than threshold means $SL = 30$ dB.

Another way of thinking in decibels is the *rule of 3dB*, which says that for every 3-dB increase in sound intensity level, the intensity (I) doubles.

Figure 13.37
A decibel meter.

EXAMPLE 13.6

Let's compare the intensity and the sound intensity levels of two sounds. According to the sound level table, the sound level of normal conversation is 60 dB. How does the intensity of this sound compare with the intensity of a sound at sound intensity level 66 dB?

Increasing a sound from 60 dB to 63 dB means the intensity has doubled, according to the rule of 3 dB. Increasing a sound from 63 dB to 66 dB means the intensity has doubled again. Therefore, the intensity must have increased 2 times 2, or four times, rising from 60 dB to 66 dB.

EXAMPLE 13.7

What is the sound intensity level of a sound with intensity 10^5 times as intense as the threshold for hearing?

$$I = 10^5 \times I_o$$
$$SL = N \times 10 \text{ dB}$$
$$SL = 5 \times 10 \text{ dB}$$
$$SL = 50 \text{ dB}$$

A sound that is 100,000 times more intense than threshold is only 50 dB!

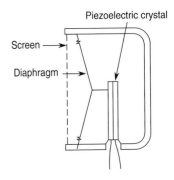

Figure 13.38
A crystal microphone. As the diaphragm moves back and forth from a sound wave, it causes the crystal to be flexed. Deforming this piezoelectric crystal produces a voltage across it proportional to the level of the sound wave.

Microphones

Most microphones in use are dynamic air pressure sensors. A diaphragm responds to a sound wave by moving back and forth. A sensor behind the diaphragm measures this motion and converts it into an electrical signal. Any type of sensor that is capable of measuring a displacement in a diaphragm can be used in a microphone. Listed below are four types of microphones (Figures 13.38, 13.39, 13.40, and 13.41).

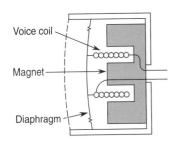

Figure 13.39
A dynamic or moving coil microphone. As the diaphragm moves back and forth from a sound wave, the attached coil moves through a magnetic field, inducing a voltage into it.

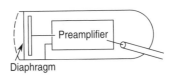

Figure 13.40
A condenser microphone measures sound levels using a capacitor. The diaphragm acts as a movable plate on a capacitor that responds to sound pressure.

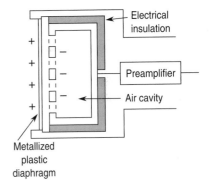

Figure 13.41
An electret-condenser microphone is essentially a condenser microphone with a permanent charge placed on the diaphragm at manufacturing. A built-in preamplifier, such as a field effect transistor (FET), amplifies the small signals.

Microphone Sensitivity

Microphone sensitivity is normally given in terms of decibels, so a reference level is often used where V is voltage and μ bar stands for microbar, which is sound pressure. For example, a microphone may have a sensitivity of −65 dB with a reference of 1V/μ bar at 1 kHz. To determine the output voltage of a microphone at various sound levels, use the following formula:

$$20 \cdot \log V = S + S_L - 74 \text{ dB}$$

Where

V = output voltage
S = sensitivity (dB)
S_L = sound level (dB)

EXAMPLE 13.8

Suppose a microphone has a sensitivity of −45 dB ref 1V/μ bar. What is the output voltage for a sound level of 50 dB?

$$S = -45 \text{ dB}$$
$$S_L = 50 \text{ dB}$$
$$V = ?$$
$$20 \cdot \log V = S + S_L - 74 \text{ dB}$$
$$20 \cdot \log V = -45 \text{ dB} + 50 \text{ dB} - 74 \text{ dB}$$
$$20 \cdot \log V = -69 \text{ dB}$$
$$V = (\text{antilog}(-69 \text{ dB}))^{\frac{1}{20}} = 0.35 \text{ mV}$$

CHAPTER SUMMARY

A wave has a length called a wavelength (λ), a rate or frequency (f) at which it moves up and down, and a speed (v) at which it travels. They are all related by the equation:

$$v = \lambda \cdot f$$

Velocity = wavelength × frequency

$$\text{Frequency} = \frac{1}{\text{period}}$$

Frequency has units of hertz (Hz), or cycles per second.

$$1 \text{ Hz} = 1 \frac{\text{cycle}}{\text{s}} = \frac{1}{\text{s}}$$

Transverse waves: Transverse waves oscillate perpendicular to the direction of travel of the wave. Light is a transverse wave.

Longitudinal waves: Longitudinal waves oscillate in the direction of travel of the wave. Sound is a longitudinal wave.

Refraction: Refraction means the bending of a wave when it moves from one medium into another.

Superposition: Many waves can travel through the same medium without affecting each other.

Constructive interference: Waves that are in step or in phase with one another add to make a bigger wave.

Destructive interference: Waves that are out of step or out of phase with one another subtract to make a smaller wave.

Tone: The tone of a sound describes the way multiple waveforms making up the sound wave act together.

Resonance: Resonance is the frequency where maximum energy transfer occurs.

Diffraction: Diffraction is the bending of waves around obstacles.

Doppler shift: Doppler shift is the shifting of the frequency of a wave because of a relative motion between the source and the observer. Mathematically, the frequency shift from the Doppler effect for light and sound is given by:

For light,

$$f_o = f_s \cdot \frac{\sqrt{1 - v_s^2/c^2}}{1 \pm v_s/C}$$

← + sign for receding sources, − sign for approaching sources

For sound,

$$f_o = f_s \cdot \left(\frac{v \pm v_o}{v \mp v_s}\right)$$

← $\frac{-}{+}$ sign for a source or observer moving away from each other

$\frac{+}{-}$ sign for a source or observer moving toward each other

Where

f_o = frequency measured by the observer,
f_s = frequency of the source

c = the speed of light,
v = the speed of sound

v_s = speed of the source,
v_0 = speed of the observer

In general, the speed of sound changes with temperature according to:

$$v = 331\frac{m}{s} + \left(0.6\frac{m/s}{°C}\right) \cdot T_c$$

Pitch: The pitch of sound is the same as the frequency.

Sound intensity: The amount of power delivered by the wave per unit area is given by:

$$I = \frac{P}{A}$$

$$\text{Intensity} = \frac{\text{power}}{\text{area}}$$

If we consider a point source of sound, then the sound radiates outward spherically, and the intensity is given by:

$$I = \frac{P}{A}$$

$$I = \frac{P}{4\pi r^2}$$

The threshold intensity for hearing is $I_o = 1 \times 10^{-12}$ W/m^2.

Sound intensity level is measured in decibels (dB) and is given by:

$$dB = 10 \cdot \log\left(\frac{I}{1 \times 10^{-12} \text{ W/m}^2}\right)$$

$$\text{Sound intensity level} = 10 \cdot \log\left(\frac{\text{intensity}}{\text{hearing intensity threshold}}\right)$$

To determine the output voltage of a microphone at various sound levels, use the following formula:

$$20 \cdot \log V = S + S_L - 74 \text{ dB}$$

Where,

V = output voltage
S = sensitivity (dB)
S_L = sound level (dB)

Problem-Solving Tip:

■ When solving problems involving the formula $v = \lambda \cdot f$, be sure to convert f into Hertz.

PROBLEMS

1. If a water wave comes onto the shore every 5 seconds, what is its frequency?

2. If you are listening to a radio station that broadcasts at 96 MHz, what is the wavelength of this wave?

3. If a signal from a satellite takes 0.01 seconds to travel to Earth, how far away is the satellite?

4. How can you create an electromagnetic wave?

5. How does a stereo speaker produce sound?

6. How are fiber-optic cables able to carry much more information than a copper wire of the same size?

7. How can many conversations take place at the same time on a phone line and not get mixed together?

8. What is the difference between a transverse wave and a longitudinal wave?

9. A police officer is shooting his radar gun in the X-band at an oncoming car traveling at 12 mi/h. What is the change in frequency of the returned signal?

10. An ultrasonic ranger and a laser ranger are used to determine the distance to an object 25 m away. How long does it take for each signal to go out and return after reflecting off this object?

11. What is the speed of sound at 28°C?

12. When woodwind instruments warm up, they become "sharp" because the air inside the instrument warms up. Why would this make the instrument sound sharp?

13. Suppose a microphone has a sensitivity of -65 dB ref $1V/\mu$ bar. What is the output voltage for a sound level of 40 dB?

14. What is the sound intensity level of a sound that is 10^5 times more intense than the threshold of hearing?

15. A 0.010-W point source of sound radiates outward equally in all directions. What is the intensity 3.0 m away from the source?

16. What is the sound intensity level of the sound in problem 15?

17. A microphone is placed 2.0 m from a 0.05-W point source of sound radiating outward equally in all directions and has a sensitivity of -50 dB ref $1V/\mu$ bar. What is the voltage output from this microphone if it is monitoring this sound?

Light

This chapter will discuss electromagnetic waves, with a focus on light and its applications. The ideas contained in this chapter should help you understand the basics behind optical storage of information, that is, compact discs (CDs), fiber optics used in telecommunications, as well as a host of other devices that use light.

ELECTROMAGNETIC WAVES

Electromagnetic waves are different than water and sound waves because they can travel through a vacuum, that is, they need no medium in which to travel. Electromagnetic waves are made of electric and magnetic fields that change in time and oscillate perpendicular to the direction of travel (Fig. 14.1).

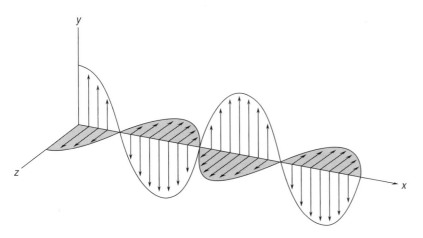

Figure 14.1
Light is a wave made of electric and magnetic fields that oscillate perpendicular to the direction of travel.

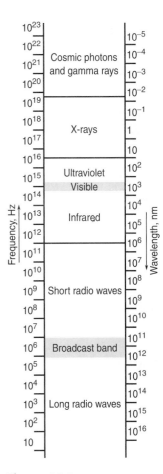

Figure 14.2
The electromagnetic spectrum.

The electromagnetic spectrum consists of all electromagnetic waves of various frequencies (Fig. 14.2). If the electromagnetic fields oscillate in the right frequency range, we can see them. Light is just part of a whole spectrum of electromagnetic waves. At other frequencies they may be radio waves, x-rays, microwaves, and so on.

THE VISIBLE SPECTRUM

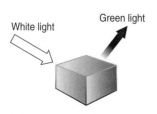

Figure 14.3
White light is composed of many different colors.

Figure 14.4
The color of an object is the result of the color of the white light shining on the object that does not get absorbed.

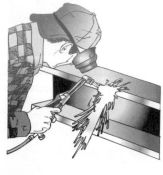

Figure 14.5
The infrared part of the spectrum is associated with the transfer of heat.

The *visible spectrum* is the part of the spectrum that we can see. Different colored lights correspond to different wavelengths of light.

White Light

White light is light composed of many different colors (Fig. 14.3). The colors can be separated out using a prism.

Colors

In order to see an object, light must be reflected off it. If you shine white light on an object and all the colors that compose the white light are absorbed except, for example, green, the object will appear green. Paints are essentially color absorbers. The color of the paint is the color from white light that does not get absorbed (Fig. 14.4).

Infrared

At wavelengths slightly longer than visible light lies the infrared part of the spectrum. *Infrared radiation* is otherwise known as *thermal radiation* or *radiant heat* (Fig. 14.5).

Ultraviolet

At wavelengths slightly shorter than visible light lies the ultraviolet part of the spectrum. Ultraviolet rays, or UV rays, cause sunburn. Welders wear goggles to protect their eyes from UV rays (Fig. 14.6).

GENERATION OF ELECTROMAGNETIC WAVES

Electromagnetic waves can be generated in different ways that lend themselves to the creation of specific parts of the spectrum.

Radio Waves

Wiggling an electric charge will generate radio waves. To understand this better, consider an electron with its electric field lines emanating from it. If a charge is wiggled up and down, these lines will be moved up and down, similar to a jump rope being snapped, generating a wave traveling down a rope (Fig. 14.7). Like the jump rope, the electromagnetic wave will be transverse, but unlike the jump rope, no medium is needed for the wave to travel on. Accompanying this wave will also be a magnetic field (this will be explained in the next chapter). The radio waves generated will have the same frequency as the frequency the electrons are wiggled at. A radio transmitter, along with its antenna, is effectively an electron wiggler.

Microwaves

Accelerating electrons in a circular path will generate microwaves. A *magnetron* is a device that generates microwaves using a magnet to accelerate a current of

Figure 14.6
Large amounts of UV light are produced in welding or cutting metals.

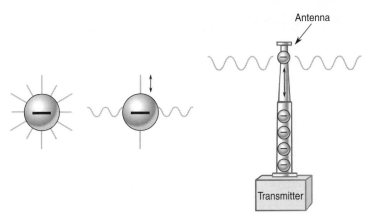

Figure 14.7
Wiggling a charge creates an electromagnetic wave. Radio transmitters wiggle electrons up and down in an antenna, producing electromagnetic waves.

Figure 14.8
Generating microwaves using a magnetron.

electrons into a circle, using a magnetic field (Fig. 14.8). Accelerating electrons like this causes them to radiate electromagnetic waves in the microwave region of the spectrum. Every microwave oven contains a magnetron.

 ## LIGHT

Electromagnetic waves also come from atoms. These waves are much shorter than radio waves. Orbiting every atom are electrons, some of which have more energy than others. If an atom is given energy, or *excited,* the electrons orbiting the atom begin jumping orbits. When an electron jumps from a high-energy orbit to a lower-energy orbit, the atom gives off an electromagnetic wave called a *photon,* or a packet of energy (Fig. 14.9). The photon has a frequency, or color, related to the energy difference between the orbits. If the energy range between the two orbits is just right, visible light will come out. *Sodium vapor lights* illustrate this clearly. The photons coming out of sodium vapor as a result of the electrons jumping orbits correspond to a wavelength of 590 nm, or yellow light.

Incandescent light bulbs generate light by heating the filament inside with an electric current. This heat energizes many different electrons in the filament in many different ways. Photons of various wavelengths will be produced, resulting in white light (Fig. 14.10). Most incandescent light bulbs emit only 10% of their energy in the visible part of the spectrum.

Fluorescent lights contain a low-pressure mercury vapor (Fig. 14.11). At one end of the tube is a filament; at the other end is an electrode at a larger voltage, compared with the first. Electrons coming off the hot filament are accelerated toward the high-voltage end, striking mercury atoms en route, which causes the mercury atom to excite then de-excite, giving off UV light. The UV light then strikes a fluorescent coating on the inside of the tube, exciting these atoms, which, in turn, give off visible light. Fluorescent lights are much more energy efficient than incandescent light, putting out about 40% of their light in the visible part of the spectrum.

Figure 14.9
Light emitted from atoms results from electrons jumping from one energy orbit to a lower one.

Lasers

Laser light is different than light from a lightbulb. Light from a lightbulb is white light, that is, of many frequencies. A laser light contains one frequency; in other words, it is monochromatic (Fig. 14.12).

Figure 14.10

Light is produced in incandescent lights by heating of the filament using an electric current. This heat is energizing so many electrons in so many different ways that the output is composed of many frequencies, or white light.

Many frequencies

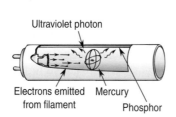

Ultraviolet photon

Electrons emitted   Mercury
from filament
            Phosphor

Figure 14.11
A fluorescent light.

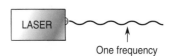

LASER

One frequency

Figure 14.12
White light contains many frequencies. Laser light contains one frequency and so is monochromatic.

LASER

Figure 14.13
White light is incoherent. Laser light is coherent.

Light waves coming out of a light bulb are all out of sync with one another. The waves are said to be incoherent. Laser light is coherent, so that all the waves of the light are in step with one another (Fig. 14.13).

The mechanism by which gas lasers and solid-state lasers work is the same. The process consists of the following steps:

Step 1   Electrons in their orbits around their atoms are excited to higher-energy orbits (Fig. 14.14). Excitation can be done in various ways using light, electrical discharge, chemical reactions, and so on.

Step 2   Excited electrons begin to jump down to lower-energy orbits. The electrons spend different amounts of time in the various orbits before jumping to the next lower orbit. If the electrons spend a lot of time in one particular orbit, soon there will be a buildup of electrons there. This is called *population inversion* because a large number of the electrons of that atom now reside at that level (Fig. 14.15).

Step 3 When electrons start to jump from a highly populated level to a lower level, light is emitted. This affects the other electrons sitting on the other atoms, causing them to jump down also. This is called *stimulated emission* (Fig. 14.16). The light emitted is all in step with the light that stimulated the emission and is of the same wavelength because the energy differences are all the same.

This whole process is contained in the acronym *laser,* which stands for *Light Amplification* by *Stimulated Emission* of *Radiation*.

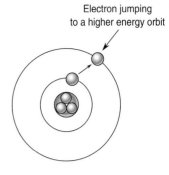

Figure 14.14
Electron excitation.

 ## THE SPEED OF ELECTROMAGNETIC WAVES

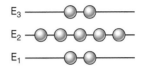

Figure 14.15
Population inversion.

Light travels at 186,000 mi/s or 3×10^8 m/s in a vacuum. A convenient conversion is that light travels 1 ft in 1 nanosecond (ns).

EXAMPLE 14.1

How long does it take for a radio signal to reach a satellite orbiting 22,000 mi above the earth (Fig. 14.17)?

$$d = 22{,}000 \text{ mi}$$

$$v = 186{,}000 \; \frac{\text{mi}}{\text{s}}$$

$$v = \frac{d}{t}$$

$$t = \frac{d}{v} \quad \leftarrow \quad \text{after some algebra}$$

$$t = \frac{22{,}000 \; \cancel{\text{mi}}}{186{,}000 \; \dfrac{\cancel{\text{mi}}}{\text{s}}} = 0.12 \text{ s}$$

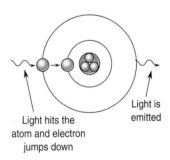

Figure 14.16
Stimulated emission.

Speed, Frequency, and Wavelength

The velocity, wavelength and frequency of any wave are given by the formula:

$$v = \lambda \cdot f$$

In the case of light,

$$v = c = 186{,}000 \text{ mi/s or } 3 \times 10^8 \text{ m/s.}$$

Therefore,

$$c = \lambda \cdot f$$

Speed of light = wavelength × frequency

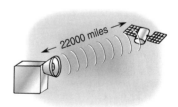

Figure 14.17

EXAMPLE 14.2

WBEZ radio station in Chicago broadcasts at 91.5 megahertz (MHz). What is the wavelength in miles and feet of this radio wave?

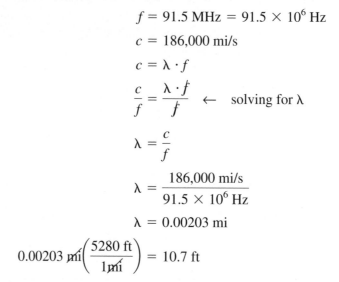

$$f = 91.5 \text{ MHz} = 91.5 \times 10^6 \text{ Hz}$$

$$c = 186,000 \text{ mi/s}$$

$$c = \lambda \cdot f$$

$$\frac{c}{f} = \frac{\lambda \cdot \cancel{f}}{\cancel{f}} \quad \leftarrow \quad \text{solving for } \lambda$$

$$\lambda = \frac{c}{f}$$

$$\lambda = \frac{186,000 \text{ mi/s}}{91.5 \times 10^6 \text{ Hz}}$$

$$\lambda = 0.00203 \text{ mi}$$

$$0.00203 \cancel{\text{mi}}\left(\frac{5280 \text{ ft}}{1 \cancel{\text{mi}}}\right) = 10.7 \text{ ft}$$

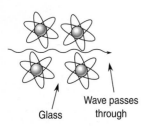

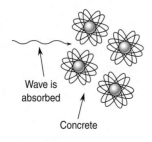

Figure 14.18
Light passes through glass and not through an opaque material because the molecules composing the glass do not absorb the light.

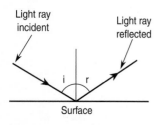

Figure 14.19
The angle of the incident ray equals the angle of the reflected ray.

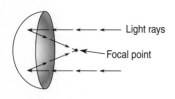

Figure 14.20
A spherical mirror.

OPTICS

Optics is concerned with generating, gathering, bending, and focusing light for some useful purpose.

Opacity

Why can you see through glass and not through a concrete wall or a nontransparent material? If a material is not transparent, it is *opaque*. Light passes through glass and not through an opaque material because the molecules composing the glass do not absorb the light (Fig. 14.18). They allow it to pass, unlike an opaque material, which absorbs the light.

Reflection of Light

Light must be reflected off an object to be able to see it. A light ray reflecting off a surface will reflect at the same angle at which it struck the surface (Fig. 14.19). If the incident and reflected angles are measured with respect to a perpendicular to the surface, then,

$$\angle i = \angle r$$

angle incident = angle reflected

Spherical Mirrors

Cut out a section of a ball, coat the inside with something reflective, and you have a spherical mirror (Fig. 14.20).

Focal Point

The focal point of a mirror is the point where all the reflected rays pass through. For a spherical mirror, the focal point is a distance from the mirror equal to half the mirror's radius (Fig. 14.21).

Parabolic Mirror

Flashlights and automobile headlights are made with parabolic reflectors (Fig. 14.22). A mirror with a parabolic shape can focus light much more sharply than a spherical mirror. Any light source at the focal point will be reflected off the mirror into a parallel of beam light. Conversely, any incoming parallel beam of light will be focused to a point. Most big telescopes are made with parabolic mirrors (Fig. 14.23).

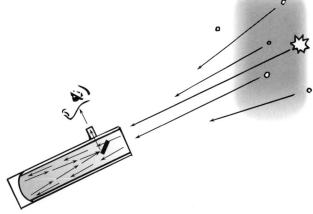

Figure 14.22
Flashlights and automobile headlights are made with parabolic reflectors.

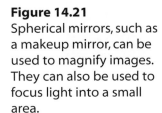

Figure 14.21
Spherical mirrors, such as a makeup mirror, can be used to magnify images. They can also be used to focus light into a small area.

Figure 14.23
Many telescopes are made with parabolic mirrors.

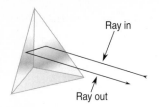

Figure 14.24
A corner reflector will take an incoming light ray and reflect it back, parallel to itself.

Corner Reflector

Coat the inside corner of a box where the sides meet the top with something reflective and you have a corner reflector (Fig. 14.24).

The astronauts left a corner reflector on the moon. It is designed so that any incoming light ray will be reflected back parallel to it (Fig. 14.25).

Figure 14.25
The distance to the moon is routinely measured by shooting a laser from earth to a corner reflector on the moon and timing how long it takes to come back.

DIFFRACTION AND REFRACTION OF LIGHT

Diffraction is the bending of waves around an obstacle (Fig. 14.26). An example of diffraction is the outline formed around shadows. The light waves bend around an obstacle and, through interference, form the outline.

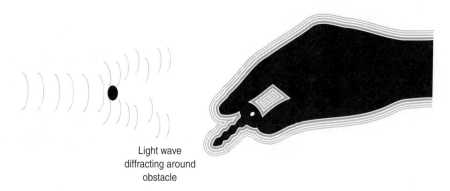

Light wave diffracting around obstacle

Figure 14.26
Light waves diffracting or bending around an obstacle.

Refraction is the bending of waves when they pass from one medium into another. An example of this is light passing from air into water. Light travels slower through water than it does through air. As a result, the wave front that enters the water will fall behind the wave in the air. The light is effectively bent because of this slowing down of the light (Fig. 14.27).

The *index of refraction (n)* is a measure of how many times slower light travels through a material than through a vacuum.

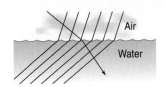

$$n = \frac{c}{v}$$

$$\text{Index of refraction} = \frac{\text{speed in a vacuum}}{\text{speed in a medium}}$$

The index of refraction is the ratio of the speed of light in a vacuum to the speed of light through some material. It is a measure of how much slower light travels in a given medium than in a vacuum.

Figure 14.27
Light is bent or refracted when moving from air to water because the wave front slows down in the water.

EXAMPLE 14.3

A light wave enters a glass lens with an index of refraction of 1.2. What is the speed of light in that lens?

$$n = 1.2$$

$$v_{\text{vacuum}} = 3 \times 10^8 \frac{m}{s}$$

$$v_{\text{glass}} = ?$$

$$n = \frac{\text{speed in a vacuum}}{\text{speed in a medium}}$$

$$\text{Speed in medium} = \frac{\text{speed in vacuum}}{n}$$

$$\text{Speed in glass} = \frac{3 \times 10^8 \frac{m}{s}}{1.2} = 2.5 \times 10^8 \frac{m}{s}$$

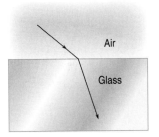

Figure 14.28

Table 14.1
Indices of refraction[†]

Material	$n = \dfrac{c}{v}$
Vacuum	1.0000
Air (at STP)	1.0003
Water	1.33
Ethyl alcohol	1.36
Glass	
Fused quartz	1.46
Crown glass	1.52
Light flint	1.58
Lucite or Plexiglas	1.51
Sodium chloride	1.53
Diamond	2.42

[†]$\lambda = 589$ nm

We can use the refractive properties of glass to make lenses bend and focus light. *Convergent lenses* are lenses that focus light (Fig. 14.29). An example is a magnifying glass.

Divergent lenses are lenses that spread out light (Fig. 14.30).

An example of a divergent lens is the eyeglass prescription for someone who is nearsighted (Fig. 14.31). The eye focuses an image short of the retina, where the image is supposed to form. The divergent lens corrects this by extending the focal length onto the retina.

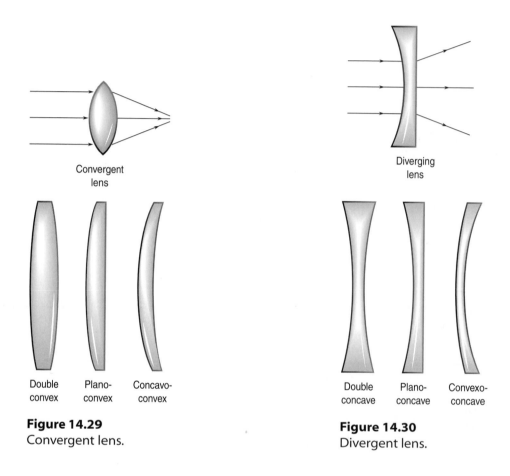

Convergent
lens

Double
convex

Plano-
convex

Concavo-
convex

Figure 14.29
Convergent lens.

Diverging
lens

Double
concave

Plano-
concave

Convexo-
concave

Figure 14.30
Divergent lens.

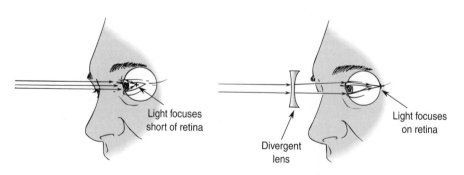

Light focuses
short of retina

Divergent
lens

Light focuses
on retina

Figure 14.31
A corrective divergent lens for nearsightedness.

Compound Lenses

Using two or more lenses together can result in much greater magnification (Figs. 14.32 and 14.33).

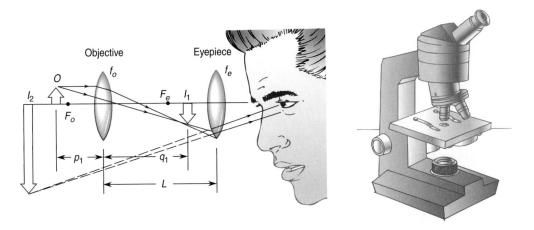

Figure 14.32
The microscope.

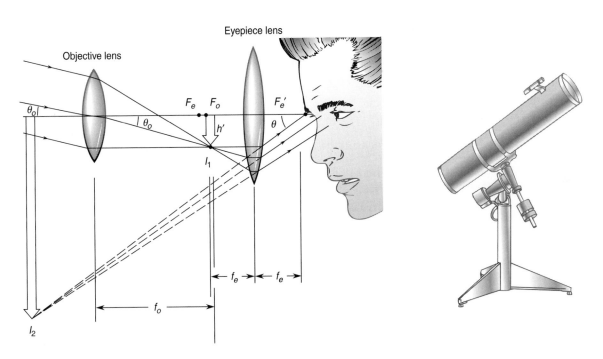

Figure 14.33
The telescope.

Aberration

Aberration is a deviation from ideal focusing of light by a lens. As a result, the image is not sharply focused. Some of the problems may lie with the geometry of the lens. Other aberrations may occur because different wavelengths of light are not refracted equally. This last problem is called *dispersion*.

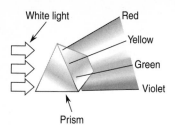

Figure 14.34
A prism separating the colors of white light.

Dispersion

Light refracts differently depending on its wavelength. A prism separates white light into colors because the speeds of the different colors of white light are different (Fig. 14.34). As a result, they are bent at different angles, with violet bending the most and red the least.

 INTENSITY AND ILLUMINATION

Luminous Intensity

The luminous intensity of a light is a measure of how much light a source emits in a given area. It is measured in candelas, a unit that is related to candles. The brightness of light sources in the past was compared to the brightness of candlelight. A typical LED puts out 40 millicandelas (mcd). Ultrabright LEDs can put out 23,000 mcd. Light emanating from an LED is concentrated in a light cone of about 30°. Another unit, the lumen (lm), is used for the intensity of a source.

For a light source that puts out light uniformly in all directions, the conversion from candelas to lumens is:

$$1 \text{ cd} = 4\pi \text{ lm} \quad \leftarrow \quad \text{For light sources that emit light in all directions uniformly}$$

A common 40-W incandescent light puts out about 500 lumens. A 40-W fluorescent light puts out about 2,300 lumens, making it much more efficient than an incandescent bulb.

Illumination

Illumination is a measure of a light's apparent brightness at some distance away from the source. The unit of illumination is $\dfrac{\text{lm}}{\text{m}^2}$ or lux.

$$1\frac{\text{lm}}{\text{m}^2} = 1 \text{ lux}$$

If the light radiates equally in all directions, the illumination is given by:

$$E = \frac{I}{4\pi r^2}$$

$$\text{Illumination} = \frac{\text{intensity (in lumens)}}{4\pi \times \text{distance}^2}$$

Figure 14.35
A light source that emits light uniformly in all directions creates a sphere of light.

The light moves outward from a source that emits uniformly in all directions, creating a sphere of light (Fig. 14.35). Think of the light as painting a sphere from the inside. The larger the sphere, the thinner the layer of light paint for a given amount of light. The light will appear to be dimmer to you because you are looking at only a small section of the sphere, painted thinly.

EXAMPLE 14.4

A light source radiates equally in all directions with an intensity of 500.0 lumens. What is the illumination 3.0 m from the source?

$$I = 500.0 \text{ lm}$$

$$r = 3.0 \text{ m}$$

$$E = ?$$

$$E = \frac{I}{4\pi r^2}$$

$$E = \frac{500.0 \text{ lm}}{4\pi(3.0 \text{ m})^2} = 4.4 \frac{\text{lm}}{\text{m}^2} = 4.4 \text{ lux}$$

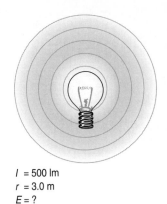

I = 500 lm
r = 3.0 m
E = ?

Figure 14.36

 ## OPTICAL DEVICES

Optical Storage

Compact discs store binary information on them in the form of bumps embedded in the plastic (Fig. 14.37). The bumps are created on a thin coating of aluminum, sandwiched between two pieces of plastic. On a CD that can store 650 megabytes of data, the bumps are 0.5 microns wide, 0.97 microns long, and 125 nm high. They are arranged in a spiral that, if stretched out, would be 5 miles long!

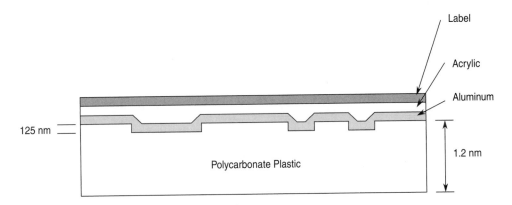

Figure 14.37
The bumps on a compact disc store binary ones and zeros.

The bumps reflect laser light that shines on them differently than on the flat areas. Electronics, along with a light sensor, measure the reflected light and interpret the information.

Digital videodiscs (DVDs) store 7.5 times more data than a CD, that is, 4.7 gigabytes. They accomplish this by making the bumps smaller and closer together, allowing them to carry more data in the same space.

Photo Diodes, Resistors, and Transistors

Resistors, diodes, and transistors can be made photosensitive and therefore activated by light (Fig. 14.38).

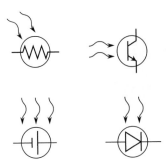

Figure 14.38
Photo sensors.

Figure 14.39
A photo-resistive cell.

Photoresistors are made from a light-sensitive material such as cadmium sulfide (Fig. 14.39). Their resistance lowers with light level (Fig. 14.40). They are used to detect light levels in cameras, streetlights, and so forth.

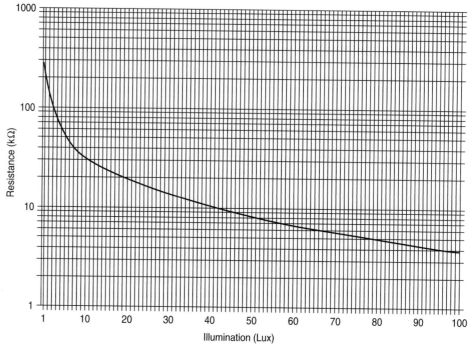

Figure 14.40
Typical response curve of a photo cell.

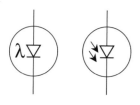

Figure 14.41
Schematic symbol of a photodiode.

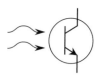

Figure 14.42
A phototransistor.

Photodiodes are essentially small solar cells in a small package (Fig. 14.41). They are used to detect light by putting out a voltage proportional to the light level.

Phototransistors have a light-sensitive base (Fig. 14.42). Instead of a wire connected to the base, a window is present, allowing light through to turn on the transistor. Phototransistors have a larger output than a photodiode due to the amplifying nature of a transistor.

Fiber Optics

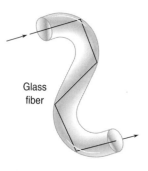

Figure 14.43
A light pipe.

A *fiber-optic* cable is a light pipe (Fig. 14.43). Light can be routed around in a light pipe similar to the way water can be routed through plumbing. Light traveling through a fiber-optic cable is internally reflected, bouncing its way through the cable until it exits at the end. A few fiber-optic cables in communications can replace big bundles of copper wire (Fig. 14.44). Signals running through copper wire dissipate heat, resulting in degradation of the signals. In fiber optics, this problem is much smaller, allowing for a much higher transmission density without degradation (Fig. 14.45).

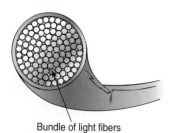

Figure 14.44
A fiber-optic cable consists of many fibers.

Solar Energy

Solar energy comes from the sun, warms us, makes plants grow, and drives the weather. The average illumination from sunlight is $1,000 \ \frac{\text{W}}{\text{m}^2}$ at the earth's surface.

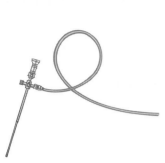

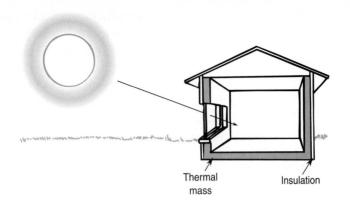

Thermal
mass

Insulation

Figure 14.45
Fiber optics is used in medical procedures, such as laproscopic surgery.

Figure 14.46

This energy can be captured and used to heat our homes or create electricity (Fig. 14.46).

Solar cells convert sunlight directly into electricity (Fig. 14.47). They are basically a very thin diode, a PN junction, with one side transparent to light. As light makes its way to the junction, it knocks free electrons on the P side, and they are transported to the N side because of the electric field present in every PN junction. A grid is placed over the solar cell to collect these electrons on the front, and a metal film is placed on the back to act as the positive terminal. The cell now acts like a battery, deriving its energy from light instead of some chemical reaction.

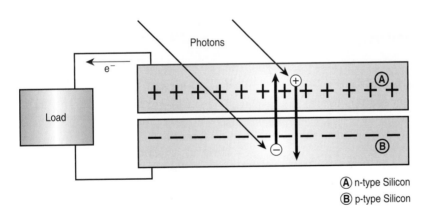

Ⓐ n-type Silicon
Ⓑ p-type Silicon

Figure 14.47
A solar cell.

The solar cell cannot convert all of the light energy into electricity because it is only about 20% efficient. Solar cells are sensitive to only a small band of wavelengths in sunlight. Seventy percent of the sunlight falling on the cell has wavelengths that are either too short or too long. Other losses are due to the high

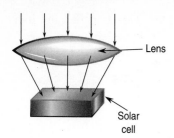

Figure 14.48
Lenses help collect more light to focus on the solar cell.

resistance of the semiconductors the cell is made of. To enhance efficiency, a cell is coated with a nonreflective agent to prevent the sunlight from just reflecting off the cell. A lens may be used to concentrate a large area of light onto a solar cell, increasing its light-gathering power (Fig. 14.48).

Currently priced at about $0.5 to $1 per kWh, solar power is much more expensive than electricity from the utility companies at about $0.09 per kWh. Only in areas where there is a lot of sunlight and where there is no electric utility is solar power becoming economically viable at this time.

Light Scattering

Our blue sky, during the day, is a result of the way sunlight scatters off of air molecules. The air molecules scatter blue light the most and red light the least. Therefore, when you look up into the sky, you see blue everywhere. The rest of the colors contained in the white light pass right on through the atmosphere. This results in the sky appearing blue (Fig. 14.49).

The reddening of the sky at night is also a result of this scattering. When the sun sets near the horizon, it must travel a greater distance through the atmosphere to reach earth than at noon. By the time the light has reached earth, all the colors have been scattered away except for red. As a result, the sky appears red (Fig. 14.50).

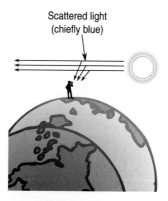

Figure 14.49
Blue light scatters more than the other colors in white light, giving the sky its blue color.

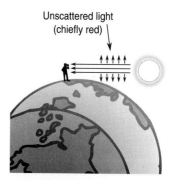

Figure 14.50
The sky appears red in the evening because light must travel a longer distance through the atmosphere, resulting in all colors except for red being scattered away.

CHAPTER SUMMARY

Light: Light is a wave made of electric and magnetic fields that change in time.

Light speed: Light travels at 186,000 mi/s or 3×10^8 m/s in a vacuum $v = \lambda \cdot f$. In the case of light $v = c =$ 186,000 mi/s or 3×10^8 m/s. Therefore,

$$c = \lambda \cdot f$$

Speed of light = wavelength $\times$ frequency

Laser: Laser stands for *L*ight *A*mplification by *S*timulated *E*mission of *R*adiation.

Luminous intensity: The luminous intensity (I) of a light is a measure of how much light a source emits in a given area. It is measured in candelas. For a light source that emits light uniformly in all directions, the conversion from candelas to lumens is:

$$1 \text{ cd} = 4\pi \text{ lm}$$ ← For light sources that emit light in all directions uniformly

Illumination is a measure of a light's apparent brightness at some distance away from the source. The unit of illumination is $\dfrac{\text{lm}}{\text{m}^2}$ or lux.

$$1\dfrac{\text{lm}}{\text{m}^2} = 1 \text{ lux}$$

If the light radiates equally in all directions, the illumination is given by:

$$E = \dfrac{I}{4\pi r^2}$$

$$\text{Illumination} = \dfrac{\text{intensity (in lumens)}}{4\pi \times \text{distance}^2}$$

Reflection of light: The angle at which a light ray reflects off a surface is equal to the angle at which it originally struck the surface. This is called the incident angle and is equal to the reflected angle:

$$\angle i = \angle r$$

angle incident = angle reflected

Diffraction is the bending of waves around an obstacle. The **index of refraction** (n) is a measure of how many times slower light travels through a material than through a vacuum.

$$n = \dfrac{c}{v}$$

$$\text{Index of refraction} = \dfrac{\text{speed in a vacuum}}{\text{speed in a medium}}$$

Convergent lenses are lenses that focus light. An example is a magnifying glass.
Divergent lenses are lenses that spread out light.
Dispersion: Dispersion is the wavelength dependence of refraction.

Problem-Solving Tip

- When solving problems dealing with frequency, wavelength, and the speed of light, remember to convert megahertz into hertz.

PROBLEMS

1. List the different colors of the visible and near-visible part of the electromagnetic spectrum and their applications.

2. Ham radio operators sometimes bounce radio signals off the moon. If the moon is about 250,000 mi away, how long does it take the signal to reach the moon?

3. How long is the wavelength of an AM radio wave of frequency 550 kHz?

4. How does an atom emit light?

5. List the various steps involved in the workings of a laser.

6. Design a streetlight that is activated at night using a photo resistor.

7. Suppose a 100-W light bulb puts out 1,250 lumens of light. What is its illumination at 2.0 m away?

8. How can you use the illumination equation to find the distance to a source of light with a known intensity?

9. Suppose you take a parabolic reflector with a collection area of 3 m^2 and place a solar cell at its focus. How many watts of sunlight will be focused onto the solar cell if the average illumination is 1,000 $\dfrac{\text{W}}{\text{m}^2}$? How many watts of electricity will be generated if the cell's efficiency is 20%?

10. How do lenses bend light?

11. What is the difference between ultraviolet and infrared light in terms of wavelength?

12. When white light shines on an object and the object appears to have a certain color, what has happened to the other colors in the white light?

13. Why can we see through glass?

14. If a quarter-wave antenna whose length is one-fourth the length of the wave is designed to work with a radio station operating at 93 MHz, how long should the antenna be?

15. If light strikes a mirror at an angle of 30° with respect to a perpendicular to the mirror, what is the angle of the reflected light?

Magnetism

Magnetism and electricity are closely related. One can be generated from the other and both have many applications. This chapter will discuss magnetism and its applications, as well as how to calculate and measure the strength of magnetic fields.

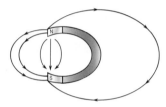

Figure 15.1

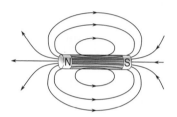

Figure 15.2
Magnetic fields emanating from a bar magnet.

Magnetic materials come about because of the way the electrons in a material interact with one another. The electrons themselves act like little magnets. If there is a net alignment of these magnetic moments, the material will have a magnetic field emanating from it (Fig. 15.2).

Materials that can be magnetized and retain their magnetism are called *permanent magnets*. Not all materials are magnetic (Fig. 15.3). For example, copper and stainless steel are nonmagnetic. Permanent magnets are usually made from iron, steel, or alloys such as alnicol (aluminum, nickel, and cobalt).

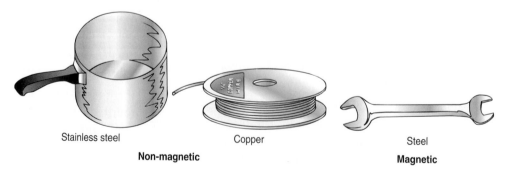

Stainless steel   Copper   Steel

Non-magnetic   **Magnetic**

Figure 15.3
Some magnetic and nonmagnetic materials.

All magnets have a north pole and a south pole. The like poles repel and the unlike poles attract one another (Fig. 15.4). This repulsion of like poles has many applications, such as magnetic bearings and magnetically levitated trains.

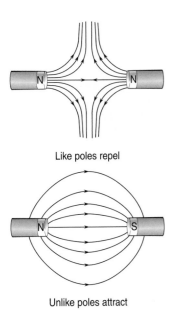

Like poles repel

Unlike poles attract

Figure 15.4
Like poles repel. Unlike poles attract.

 ## ELECTROMAGNETS

Current moving through a wire will produce a magnetic field (Fig. 15.5). Passing a current through a coil of wire produces a magnetic field similar to a field from a bar magnet. By turning the current on or off, the field can be turned on or off (Fig. 15.6).

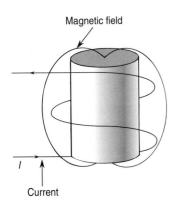

Figure 15.5
An electromagnet, magnetism generated by an electric current.

Figure 15.6
An electromagnet used at a junkyard.

Relay

A relay is an electromagnetic switch that consists of a coil and a switch (Fig. 15.7). When a current passes through the coil, it generates a magnetic field, which pulls the switch closed.

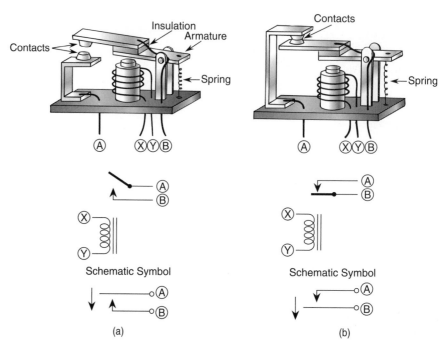

Figure 15.7
A relay is an electromagnetic switch.

Superconducting Magnets

Generating magnetic fields using standard electromagnets is limited because the field strength is proportional to the current. The higher the current, the greater the heating of the wire, $P = I^2 \cdot R$. Big electromagnets need to be cooled with water. To obtain higher magnetic fields, a superconducting wire is used instead of copper wire. Superconducting wire generates virtually no heat. The only problem with it is that it has to be cooled to very low temperatures using liquid helium and it requires a costly cryogenic system (Fig. 15.8).

Magnetically Levitated Trains

A magnetically levitated train consists of superconducting magnets placed on the bottom of the train and normal coils placed in the track (Fig. 15.9). When the train passes over the coils in the track, a current is induced in these coils. This current generates a magnetic field that opposes the superconducting magnets' field. The like poles of the two magnets repel, and the train floats.

Figure 15.8
A superconducting magnet.

EARTH'S MAGNETIC FIELD

Inside the earth there is molten iron. In this state, the iron is like a soup, with atoms missing some of their electrons together with those electrons moving freely about.

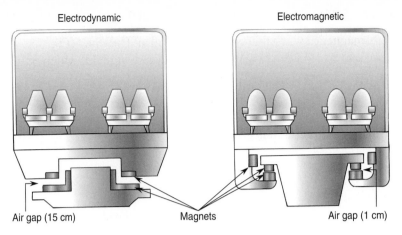

Electrodynamic

Electromagnetic

Air gap (15 cm)

Magnets

Air gap (1 cm)

Figure 15.9
A magnetically levitated train.

Because the earth is spinning, so is the molten iron. A magnetic field is generated from this rotating mixture similar to that of an electromagnet (Fig. 15.10).

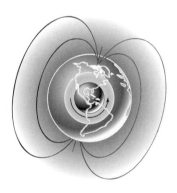

Figure 15.10
Earth's magnetic field is due to electric currents in the molten iron flow near its core.

Figure 15.11
A compass aligning with the earth's magnetic field.

Compass

A compass measures the direction of the earth's magnetic field. Like poles of a magnet repel, and unlike poles attract. The compass's magnetic south pole will be attracted to the earth's magnetic north pole and will rotate itself into alignment (Fig. 15.11).

 ## MAGNETIC FIELD GENERATED BY A CURRENT-CARRYING STRAIGHT WIRE

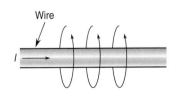

Figure 15.12
A magnetic field is generated by passing a current through a wire.

As stated earlier, passing current through a wire will generate a magnetic field. The magnetic field will surround the wire as shown in Figure 15.12.

The unit for magnetic field is the tesla. The earth's magnetic field has a strength of approximately 5.2×10^{-5} tesla. Another unit of magnetic field is the gauss.

$$1 \text{ tesla (T)} = 10^4 \text{ gauss}$$

For a straight wire, the field strength is given by:

$$B = \frac{\mu_0 \cdot I}{2\pi r}$$

$$\text{Magnetic field (tesla)} = \frac{4\pi \times 10^{-7} \times \text{current (amps)}}{2\pi \times \text{distance (meters)}}$$

where μ_0 is called the permeability and is given by $\mu_0 = 4\pi \times 10^{-7}\frac{\text{T} \cdot \text{m}}{\text{A}}$

EXAMPLE 15.1

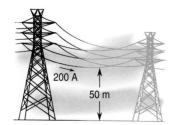

Figure 15.13

What is the strength of a magnetic field 50 m from a power line carrying 200 A of current?

$$I = 200 \text{ A}$$

$$r = 50 \text{ m}$$

$$B = ?$$

$$B = \frac{4\pi \times 10^{-7} \cdot I}{2\pi r}$$

$$B = \frac{4\pi \times 10^{-7} \cdot 200 \text{ A}}{2\pi \cdot 50 \text{ m}} = 800 \times 10^{-9}\text{T}$$

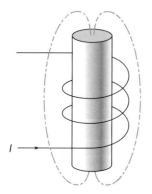

Figure 15.14
A magnetic field generated by an electromagnet.

MAGNETIC FIELD INSIDE A COIL

Passing a current through a coil of wire produces a magnetic field similar to a field from a bar magnet (Fig. 15.14).

For a long, tightly turned coil of length L with N turns, the magnetic field is given by:

$$B = \mu_0 I\left(\frac{N}{L}\right)$$

$$\text{Magnetic field (tesla)} = \left(4\pi \times 10^{-7}\frac{\text{T} \cdot \text{m}}{\text{A}}\right)$$

$$\times \text{ current (amps)} \times \left(\frac{\text{turns}}{\text{length (meters)}}\right)$$

EXAMPLE 15.2

A current of 5.0 A passes through a coil of length 0.10 m with 4,500 turns. What is the strength of the field inside the coil?

$$I = 5.0 \text{ A}$$

$$L = 0.10 \text{ m}$$

$$N = 4,500 \text{ turns}$$

$$B = ?$$

$$B = \mu_0 I \left(\frac{N}{L}\right)$$

$$B = 4\pi \times 10^{-7}\frac{\text{Tm}}{\text{A}} \cdot 5.0 \text{ A} \cdot \left(\frac{4,500}{0.10 \text{ m}}\right) = 0.28 \text{ T}$$

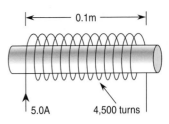

Figure 15.15

ELECTROMAGNETIC INDUCTION

Electromagnetic induction is the generation of electricity from magnetism. Magnetic fields can push electrons through a conductor just like a battery can. In order for the electricity to be generated, however, the magnetic field must be changing in some way, near the wire (Fig. 15.16). Rotating a coil or wire near the magnet, or vice versa, will accomplish this.

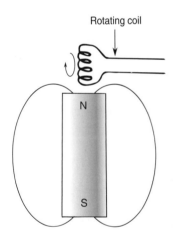

Figure 15.16
A changing magnetic field will generate a current in the coil.

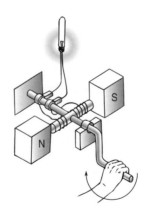

Figure 15.17
An electrical generator

AN ELECTRICAL GENERATOR

A simple electrical generator consists of a rotating coil in a magnetic field (Fig. 15.17). As the coil rotates, the penetration of the magnetic field through the coil changes. By induction, this changing magnetic field through the coil induces an AC current in it.

ALTERNATORS

An automobile alternator consists of a rotating coil inside a set of three fixed coils (Fig. 15.18). The rotating coil is connected to a source of current, the car battery, and therefore generates a magnetic field. This rotating magnetic field induces an AC current into the fixed coils. The AC output is then rectified and fed to the battery.

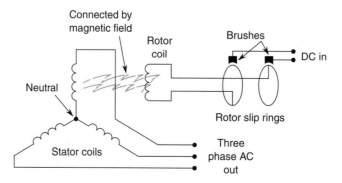

Figure 15.18
An automobile alternator.

 MOTORS

DC Motors

A permanent-magnet DC electric motor consists of a rotating coil with current running through it, centered around a permanent magnet. Current running through the coil produces its own magnetic field, which creates an attraction to the fixed magnet, causing it to spin (Fig. 15.19).

The current through the coil is made to change direction halfway through a rotation to keep it spinning. Otherwise, it would get stuck in one place. The rotating coil is called the *rotor,* the permanent magnet is called the *stator,* and the device that changes the current's direction is called the *commutator* (Fig. 15.20). Current is delivered by a set of brushes that rub the commutator.

Most DC motors have three permanent magnet pole pairs (Fig. 15.21). Having more than one pole pair keeps the motor from getting stuck in one position and keeps the rotation smoother.

A DC motor can act as a generator (Fig. 15.22 on p. 228). Spinning the motor's shaft will generate a voltage.

AC Motors

The most commonly used AC motor is the AC induction motor, which has no brushes or commutator. The stator consists of coils with AC current passing through them, producing a rotating magnetic field (Fig. 15.23 on p. 228). The rotor has a squirrel cage–type construction (Fig. 15.24 on p. 228). As the magnetic field rotates around the squirrel cage, a current is induced into the coil below the field. This, in turn, produces its own magnetic field and generates a torque with the stator's magnetic field, causing the rotor to rotate (Fig. 15.25 on p. 229). AC motors can operate with either a clockwise or a counterclockwise rotation.

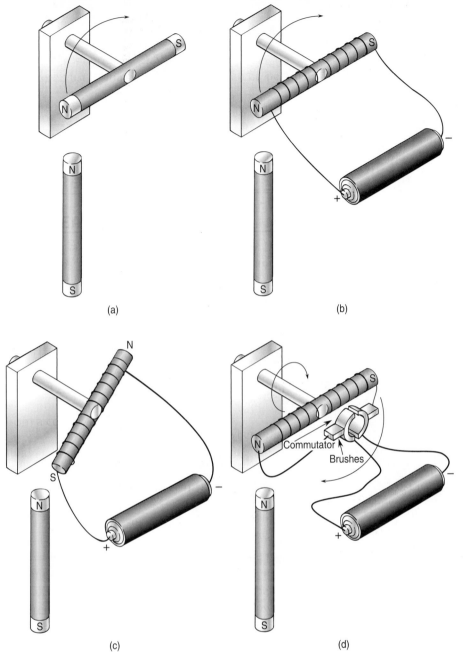

(a)   (b)

(c)   (d)

Figure 15.19
The interaction of the two fields results in the coil rotating into alignment.

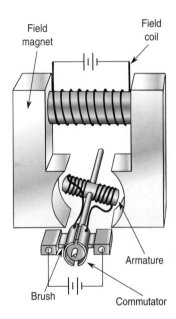

Figure 15.20
The direction of the current through the rotor changes with the aid of the commutator.

Figure 15.21
A six-pole DC motor.

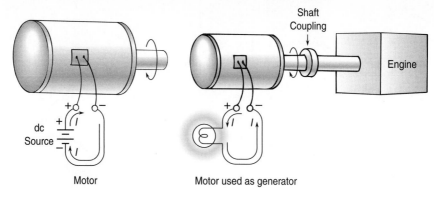

Figure 15.22
A DC motor used to generate electrical power.

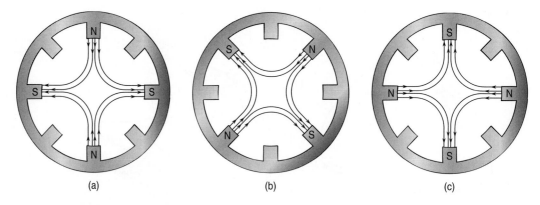

Figure 15.23
The stator produces a rotating magnetic field.

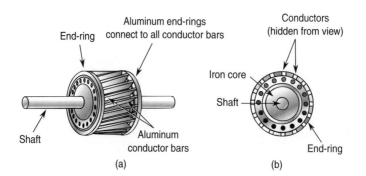

Figure 15.24
The rotor is made like a squirrel cage.

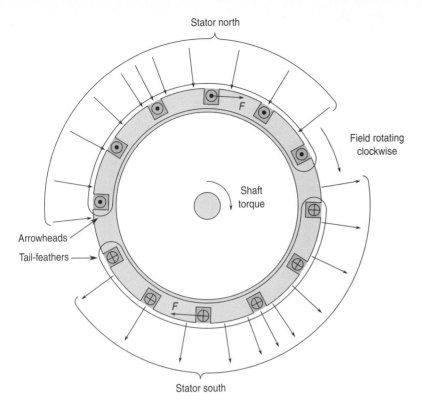

Figure. 15.25
When a magnetic field passes through one of the coils, a current and magnetic field are created, interacting with the applied field. The two fields will try to align, resulting in the rotor rotating.

 ## ELECTRIC METERS

The electric meter on your house is an energy meter that measures the number of kilowatt-hours used (Fig. 15.26). It works on the same principle as an AC induction motor. There are two electromagnets, one connected across the source voltage to the house, the other connected in series with source. Between the two electromagnets is an aluminum disk that is free to rotate. Because of the way the two electromagnets are connected, they are out of phase with one another by 90° and produce a rotating magnetic field. These electromagnets induce currents into the aluminum disk, which produces a magnetic field. This magnetic field interacts with the electromagnets' field, causing the disk to rotate. The number of rotations the motor makes is proportional to the current, voltage, and the time the motor is running (recall $E = I \times V \times t$). A counting mechanism records this number in terms of kilowatt-hours (kWh).

 ## MAGNETIC MEMORY

Magnetic Tape

Tape recorders record and play back information stored on magnetic tape. The magnetic tape is made by applying a mixture of ferric oxide powder and glue to a piece of plastic tape. Iron oxide (FeO), is common rust. Ferric oxide (Fe_2O_3) is

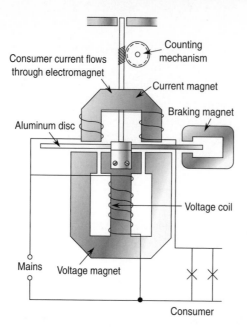

Figure 15.26
An electric meter or an energy meter.

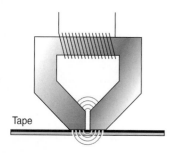

Figure 15.27
A tape recorder's head storing its magnetic field on the magnetic tape.

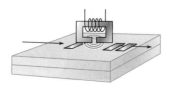

Figure 15.28
The original electrical signal that magnetized the tape is now re-created during playback.

another oxide of iron. Both are *ferromagnetic* materials that, when exposed to a magnetic field, will remain magnetized. The magnetic tape is passed over an electromagnet called a *head.* An electrical signal from a microphone is sent into the head, generating a magnetic field. The ferric oxide in the tape lines up in size and direction with the magnetic field of the head when it passes under it. The tape therefore stores the electrical signal applied to the head, in the form of magnetism (Fig. 15.27).

To play back what was recorded, the head is used as a magnetic field sensor. As the tape is passed over the head, the magnetic field in the tape induces a current into the head. The original electrical signal that magnetized the tape is now re-created (Fig. 15.28). This signal is amplified and then sent to a speaker.

An actual cassette player contains two small heads for the left and the right channels of the stereo (Fig. 15.29). The "A" and "B" sides of the tape lie next to each other. The two heads move together to play back either side of the tape.

Side B left channel
Side B right channel
Side A right channel
Side A left channel

Figure 15.29
An actual cassette tape.

Figure 15.30
A hard disk drive.

Hard Disks

A hard disk on a computer stores information in a similar way as a cassette tape (Fig. 15.30). A hard disk, however, is much faster and can store much more information than a tape. It accomplishes this by storing the information in much smaller magnetic domains and on a high-speed rotating disk. The disk rotates at about 3,000 in/s compared with 2 in/s with tape. The head moves radially across the disk, enabling it to reach and retrieve any stored information at a very high speed.

 ## TRANSFORMERS

A transformer is a device that transforms a small AC voltage into a large one or a large voltage into a small one (Fig. 15.31–15.33). It consists of two coils next to one another. When current passes though the first coil, it generates a magnetic field, which passes through the second coil. The magnetic field pushes on the electrons in the second coil, thus producing the current.

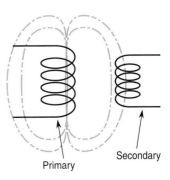

Figure 15.31
A transformer.

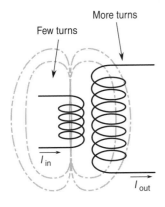

Figure 15.32
A step-up transformer.

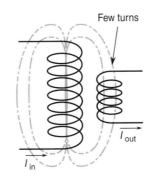

Figure 15.33
A step-down transformer.

 ## INDUCTIVE HEATING

Current running through a wire produces heat. If a varying magnetic field is placed below a cooking pot, electromagnetic induction will generate a current in the pot, causing it to heat up (Fig. 15.34).

Figure 15.34
Inductive heating.

 ## MAGNETIC RESONANCE IMAGING

Magnetic resonance imaging (MRI) is a noninvasive technique for "looking" inside a body without going inside, or sending something through the body like an x-ray (Fig. 15.35).

MRI works on the principle that different body tissues respond differently to varying magnetic fields. By measuring these different responses over an area, a two- or three-dimensional image can be formed.

Let's look in detail at how MRI works. When an atom is placed in a strong constant magnetic field, its nucleons act like small magnets themselves and will line up either parallel or antiparallel to the field (Fig. 15.36).

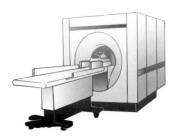

Figure 15.35
An MRI machine.

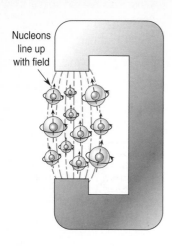

Nucleons line up with field

Figure 15.36
Nucleons in an atom line up parallel or antiparallel to the applied field.

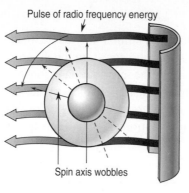

Pulse of radio frequency energy

Spin axis wobbles

Pulse of radio frequency causes spin axes to wobble

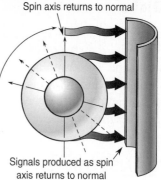

Spin axis returns to normal

Signals produced as spin axis returns to normal

Spin axis sends out signals as it returns to normal

Figure 15.37
Determining the resonance frequency of the nucleons.

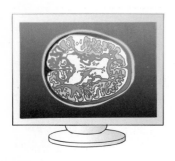

Figure 15.38
An MRI image.

In addition to the constant field, a small varying magnetic field is placed perpendicular to the constant field. This will cause the nucleons to slightly precess around their original direction. At just the right frequency, the precession will be large and will absorb a lot of energy from the varying magnetic field. This is the *resonant frequency*. The resonant frequency is different for different materials being studied, and this is how we can distinguish among materials (Fig. 15.37).

Inside the MRI machine is a large superconducting magnet of 0.5 to 2 tesla in strength, smaller field-shaping magnets, and RF (radio frequency) coils that generate a time-varying magnetic field. The large magnet does the overall aligning of the nucleons in the patient. The shaping magnets select the area of the body to scan, and the RF coils sweep through frequencies, measuring the resonance frequencies of the tissue under inspection. All this information is collected by a computer and put together to form an image (Fig. 15.38).

MEASURING MAGNETIC FIELDS

Hall Probe

A Hall probe is the most common method of measuring magnetic fields. A Hall probe consists of a semiconductor material with a current running through it. If a magnetic field penetrates the Hall probe, the current flowing inside the probe will

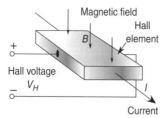

Magnetic field

Hall element

B

+

Hall voltage
V_H

−

I

Current

Figure 15.39
A Hall probe responds to a magnetic field by generating a voltage proportional to the field strength.

be pushed to one side, as shown. This will produce a voltage across the probe because the charge will be concentrated on one side. This voltage is proportional to the magnetic field strength (Fig. 15.39).

Metal Detectors

A simple metal detector consists of a resonant frequency RLC (Resistor, Inductor, Capacitor) circuit (Fig. 15.40). As the circuit is brought near some metal, the inductance of the inductor will change and therefore so will the resonant frequency. This frequency can be monitored in some way with a meter or headphones.

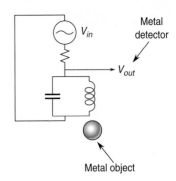

Figure 15.40
A simple metal detector consists of a resonant frequency RLC circuit.

CHAPTER SUMMARY

Materials that can be magnetized and retain their magnetism are called *permanent magnets.*

Electromagnets: Current moving through a wire will produce a magnetic field. Passing a current through a coil of wire produces a magnetic field similar to a field from a bar magnet.

Relay: A *relay* is an electromagnetic switch that consists of a coil and a switch. When a current passes through the coil, it generates a magnetic field, which pulls the switch closed.

The unit for magnetic field is the tesla. Another unit of magnetic field is the gauss.

1 tesla (T) = 10^4 gauss

For a straight wire the field strength is given by:

$$B = \frac{\mu_0 \cdot I}{2\pi r}$$

Magnetic field (tesla) =
$$\frac{4\pi \times 10^{-7} \times \text{current (amps)}}{2\pi \times \text{distance (meters)}}$$

where μ_0 is called the permeability and is given by
$$\mu_0 = 4\pi \times 10^{-7}\frac{\text{T} \cdot \text{m}}{\text{A}}$$

Magnetic field inside a coil: Passing a current through a coil of wire produces a magnetic field similar to a field from a bar magnet.

For a long, tightly turned coil of length L with N turns, the magnetic field is given by:

$$B = \mu_0 I \left(\frac{N}{L}\right)$$

Magnetic field (tesla) = $\left(4\pi \times 10^{-7}\frac{\text{T} \cdot \text{m}}{\text{A}}\right)$

$\times$ current (amps) $\times \left(\frac{\text{turns}}{\text{length (meters)}}\right)$

Electromagnetic induction means generating electricity from magnetism.

DC motors: A DC electric motor consists of a rotating coil with current running through it centered around a permanent magnet. Current running through the coil produces its own magnetic field, which creates an attraction to the fixed magnet, causing it to spin.

AC motors: The most commonly used AC motor is the AC induction motor. It has no brushes or commutators.

Magnetic tape: Tape recorders record and play back information stored on magnetic tape. The magnetic tape is made by applying a mixture of ferric oxide powder and glue to a piece of plastic tape.

Ferromagnetic materials, when exposed to a magnetic field, will remain magnetized.

Problem-Solving Tip

- There are only two formulas for this chapter, making it easy to choose which formula to use. The formula for the magnetic field at the center of the coil has turns in the formula, and the other is for a straight wire.

PROBLEMS

1. What is the origin of the earth's magnetic field, and how would you measure it?

2. Explain how electricity can be generated from magnetism and vice versa.

3. Two magnets with like poles facing one another will repel. How can this fact be used to build a magnetic bearing?

4. What is a relay, and how does it work?

5. What is the strength of a magnetic field 1.0 m from a wire carrying a current of 20.0 A?

6. A reed switch consists of a glass tube containing a switch made from two pieces of spring metal separated from each other. When a magnet is brought near, the field pulls the switch closed. How can this be used as a proximity switch?

7. What is the magnetic field inside a long, thin coil of length 0.08 m, 2,300 turns, with a current of 0.50 A flowing through it?

8. How does the motor work in a toy slot car?

9. Suppose you build an electrical generator from a car alternator powered by a 3-hp lawn mower engine. If the efficiency of the alternator is 90% and the alternator is putting out 12 V, how much current can it deliver? Note 1 hp = 746 W.

10. If a Hall probe has a sensitivity of 22 gauss/mV and is putting out 5,000 mV, what is the field strength?

11. How does the electric meter on your house measure the electrical energy you use?

12. How does a magnetic tape, floppy disk, or credit card work, and why is it a bad idea to bring magnets near them?

13. Design a windmill that generates electricity from a DC motor and store it in a battery.

14. What is the magnetic field at the ground 6.0 m below a power line carrying 350 A?

15. Suppose you want to test the effect of magnetic fields on a gerbil. To generate the field, you make a coil from a toilet paper roll measuring 0.025 m in radius and wrap 500 turns of wire around it. If the wire has a resistance of 1.5 Ω and is connected to a 12-V battery, what is the magnetic field on the gerbil?

Trigonometry Review

Trigonometry shows us the relationship between the sides of a triangle and its angles. It is very useful in solving vector problems in two or three dimensions. In electronics, trigonometric functions are used because AC signals repeat themselves in the same way that these functions repeat themselves periodically.

A right triangle is a triangle with a 90° angle.
Key points about a right triangle:

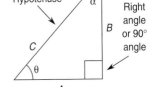

Fig. A1

- The long side of the triangle is called the *hypotenuse*.

- $\alpha + \theta = 90°$

- **Pythagorean theorem:** The hypotenuse equals the square root of the sum of the squares of the sides.

$$C = \sqrt{A^2 + B^2}$$
$$A = \sqrt{C^2 - B^2}$$
$$B = \sqrt{C^2 - A^2}$$

EXAMPLE A.1

Find the length of the sides of the following triangles: a) $A = 3, B = 4, C = ?$; b) $A = 7, C = 9, B = ?$; c) $B = 13, C = 18, A = ?$

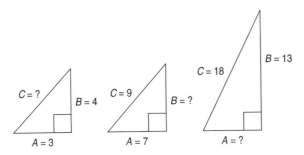

Fig. A2

a) $A = 3.0, B = 4.0, C = ?$
$$C = \sqrt{A^2 + B^2}$$
$$C = \sqrt{3.0^2 + 4.0^2}$$
$$C = 5.0$$

b) $A = 7.0, C = 9.0, B = ?$

$$B = \sqrt{C^2 - A^2}$$
$$B = \sqrt{9.0^2 - 7.0^2}$$
$$B = 5.7$$

c) $B = 13.0, C = 18.0, A = ?$

$$A = \sqrt{C^2 - B^2}$$
$$A = \sqrt{18.0^2 - 13.0^2}$$
$$A = 12.4$$

The hardest part of trigonometry is determining what side is opposite or adjacent to an angle.

- The *opposite side* of an angle can be found by drawing a line through the angle. This line will strike the opposite side.
- The *hypotenuse* is the long side of the triangle. It is the side opposite the 90° angle.
- The *adjacent side* is the side next to an angle. The hypotenuse may be next to an angle, but it is never called the adjacent side. It is always the hypotenuse.

Fig. A3

C is the hypotenuse.

B is the side opposite angle θ.

A is the side adjacent to angle θ.

A is the side opposite angle α.

B is the side adjacent to angle α.

sin(θ), cos(θ), tan(θ)

sin, cos, and tan are defined to be:

$$\sin(\theta) = \frac{\text{opposite}}{\text{hypotenuse}}$$

$$\cos(\theta) = \frac{\text{adjacent}}{\text{hypotenuse}}$$

$$\tan(\theta) = \frac{\text{opposite}}{\text{adjacent}}$$

An easy way to remember these is with the mnemonic:

SOH-CAH-TOA

Where SOH stands for sin = op/hy

Where CAH stands for cos = adj/hy

Where TOA stands for tan = op/adj

EXAMPLE   A.2

Determine the sin(θ), cos(θ), tan(θ) and sin(α), cos(α), and tan(α) of the following triangle.

Fig. A4

$$\sin(\theta) = \frac{3}{5}$$

$$\cos(\theta) = \frac{4}{5}$$

$$\tan(\theta) = \frac{3}{4}$$

$$\sin(\alpha) = \frac{4}{5}$$

$$\cos(\alpha) = \frac{3}{5}$$

$$\tan(\alpha) = \frac{4}{3}$$

By knowing the ratio of the sides $\dfrac{\text{opposite}}{\text{hypotenuse}}$, $\dfrac{\text{adjacent}}{\text{hypotenuse}}$ or $\dfrac{\text{opposite}}{\text{adjacent}}$ you can determine the corresponding angle.

$$\theta = \sin^{-1}\left(\frac{\text{opposite}}{\text{hypotenuse}}\right)$$

$$\theta = \cos^{-1}\left(\frac{\text{adjacent}}{\text{hypotenuse}}\right)$$

$$\theta = \tan^{-1}\left(\frac{\text{opposite}}{\text{adjacent}}\right)$$

EXAMPLE   A.3

Given $A = 12.0$ cm and $B = 14.0$ cm, what are θ and beta (β)?

$$A = 12.0 \text{ cm}$$

$$B = 14.0 \text{ cm}$$

$$\theta = \tan^{-1}\left(\frac{\text{opposite}}{\text{adjacent}}\right)$$

$$\theta = \tan^{-1}\left(\frac{14.0 \text{ cm}}{12.0 \text{ cm}}\right)$$

$$\theta = 49°$$

$$\beta + \alpha = 90°$$

$$\beta = 90° - \alpha$$

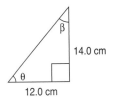

Fig. A5

$$\beta = 90° - 49°$$

$$\beta = 41°$$

Given one side and one angle of a right triangle, you can determine all other sides and angles of the triangle.

EXAMPLE A.4

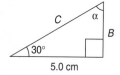

Fig. A6

Given $A = 5.0$ cm and $\theta = 30°$, what are a) $\alpha = $?; b) $B = $?; c) $C = $?

$$A = 5.0 \text{ cm}$$

$$\theta = 30°$$

a) $\alpha = $?

$$\alpha + \theta = 90°$$

$$\alpha = 90° - 30°$$

$$\alpha = 60°$$

b) $B = $?

Let's write all three functions down to help determine which one to use.

$$\sin(30°) = \frac{5.0 \text{ cm}}{C}$$

$$\cos(30°) = \frac{B}{C}$$

$$\tan(30°) = \frac{5.0 \text{ cm}}{B} \quad \leftarrow \quad \text{This one will get us } B.$$

$$B = \frac{5.0 \text{ cm}}{\tan(30°)}$$

$$B = 8.7 \text{ cm}$$

c) $C = $?

$$\sin(30°) = \frac{5.0 \text{ cm}}{C} \quad \leftarrow \quad \text{This one will get us } C.$$

$$C = \frac{5.0 \text{ cm}}{\sin(30°)}$$

$$C = 1\bar{0} \text{ cm}$$

EXAMPLE A.5

Suppose you determine the height of a building, using trigonometry. What is the height of a building when you are 1.0 mi away and measure an angle of 10.0° from the sidewalk to the top of the building?

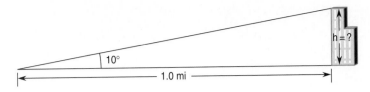

Fig. A7

$$\theta = 10°$$

$$B = 1.0 \text{ mi}$$

$$h = ?$$

$$\sin(10°) = \frac{h}{C}$$

$$\cos(10°) = \frac{1.0 \text{ mi}}{C}$$

$$\tan(10°) = \frac{h}{1.0 \text{ mi}} \quad \leftarrow \quad \text{This one will get us } h.$$

$$h = 1.0 \text{ mi} \cdot \tan(10°)$$

$$h = 0.18 \text{ mi}$$

Problem-Solving Tips

■ Make sure your calculator is in the degrees mode and not radians or gradians before calculating.

■ If you have trouble determining whether to use sin, cos, or tan, write all three down. Having it on paper is easier than solving it in your head.

PROBLEMS

1. Determine the length of the hypotenuse of a right triangle with sides of 2.3 cm and 4.6 cm.

2. What are the angles that make up the triangle in problem 1?

3. Determine the length of the side of a right triangle with a side of 14.7 in and a hypotenuse of 23.2 in.

4. What are the angles that make up the triangle in problem 3?

5. What is the height of a tree when you are 500.0 ft away and measure an angle of 25.0° from the ground to the top of the tree?

6. How long is the diagonal of a square with sides that are 6.0 cm long?

7. If an electronic signal is described by the formula $V = 2.0 \text{ V} \sin(2\pi f \cdot t)$ and $f = 1{,}254$ Hz, what is the value of V at t = 12 ms?

8. If a force of 10.0 N acts downward at a 45° angle, what component of force acts downward and horizontal?

9. At what angle are the sides of a right triangle equal?

10. At what angle is one side of a right triangle twice as big as the other?

Sensors

This appendix is a summary of the sensors covered throughout the book. The original figure numbers have been included for easy reference.

Resistors, capacitors, and inductors can be used as sensors to measure quantities such as temperature, force, and pressure. Anything that changes a resistor, capacitor, or inductor's value can be used to measure a physical quantity.

Measuring Distance

Fig. 2.7   Electronic micrometers and calipers contain a resistive strip. As the barrel or caliper is moved back, the distance between measuring points on the wire increases, and therefore the resistance measured will increase. This resistance is converted into a distance and displayed.

Fig. 2.9   The ultrasonic tape measure works by sending a sound wave and timing how long it takes to return after bouncing off an object whose distance is being measured. The longer it takes, the farther away the object.

Fig. 2.11   Laser ranging works by sending a light beam and timing how long it takes to return after bouncing off the object being measured. The longer it takes, the farther away the object. Background light does not interfere with the measurement because the detector is sensitive only to the frequency of the laser.

Fig. 2.12   A variable capacitor is used to measure displacement.

Fig. 2.14   A variable inductor is used to measure length. As the core is moved out, the inductance will decrease.

Measuring Angles

Fig. 6.4   A potentiometer used to measure angular orientation.

Fig. 6.5   A computer mouse converts the revolutions of the ball in the mouse to a position on the computer screen.

Measuring Time

A clock is used to measure time.

Fig. 2.25   A 555 timer chip.

Fig. 2.26   An LC oscillator.

Fig. 2.27   A crystal oscillator.

Fig. 2.28   An atomic clock.

Measuring Velocity

Fig. 13.23   A *radar gun* consists of a microwave radio source emitting a beam of radio waves at a given frequency. The frequency of radio waves increases after bouncing off oncoming objects. This is because the wave becomes compressed. The reflected wave returns to the gun, and its frequency is compared to the frequency of the original wave; the difference in the frequency of the two waves is a measure of how fast the car is moving.

Fig. 13.25   The Doppler effect can be applied to ultrasonic waves to determine speed. The speed of an object can be determined by bouncing an ultrasonic sound wave off of a moving object, measuring the change in frequency between the outgoing and incoming waves.

Measuring Angular Velocity

Fig. 6.12   Magnetic pickup coil tachometer. A coil is placed near a rotating gear. The inductance of the coil will change as a tooth on the gear positions itself directly below the coil. This change in inductance will be monitored and converted into a pulse and counted for some interval of time. Knowing the number of teeth on the gear and the number of pulses occurring in some interval of time, one can determine the angular velocity.

Measuring Acceleration

If something is accelerated it is because there is a force on it. Remember $a = \dfrac{F}{m}$.

We can measure the acceleration of an object by measuring the force applied to a sensor sitting on the object.

Fig. 3.19   A resistive accelerometer is essentially a strain gauge with a mass mounted on it. During acceleration, the mass produces a stress on the resistive element. Stressing the resistor produces a change in its resistance. Pendulous type accelerometers have a suspended mass with some type of transducer measuring the position of the mass. The mass will move only if accelerated while the transducer will pick up this motion.

Fig. 3.17   If a pendulum is accelerated, the pendulum will swing backward. If the pendulum is magnetic and we place a coil near it, and if the pendulum moves relative to the coil, a voltage will be induced in the coil. A voltage present on the coil will indicate acceleration.

Measuring Force

Electronic scales typically use a strain gauge to measure weight or force. Most strain gauges are resistive. A resistive strain gauge consists of a very thin resistor mounted on a material that will flex as force is applied. A force applied to the resistor will deform it slightly, resulting in a change in its resistance.

Fig. 4.5   A resistive strain gauge converts an applied force, which deforms the gauge, into a change in resistance. The resistance is then converted into the corresponding weight.

Measuring Pressure

Fig. 9.17   A typical tire gauge consists of a piston contained in a cylinder constrained by a spring. The piston is connected to a calibrated rod from which the pressure is read.

An electronic pressure gauge typically uses a strain gauge to measure pressure. A strain gauge is a device that, when strained or made to change shape, will change its electrical properties. In a pressure gauge, the strain gauge is mounted in some type of chamber that can be attached to the liquid or gas volume being measured. As the liquid or gas molecules strike the gauge, they deliver a force to the resistive pattern the sensor is made of. This will cause it to deform, resulting in a change in resistance.

Fig. 9.18   A resistive force sensor can be used to measure pressure. Recall that $P = \dfrac{F}{A}$. Therefore, a change in pressure will result in a change in resistance.

Measuring Electrical Energy

The electric meter on your house is an energy meter that measures the number of kilowatt-hours used. It works on the same principle as an AC induction motor. There are two electromagnets, connected from the source voltage to the house, the other connected in series with source to the house. Between the two electromagnets is an aluminum disk that is free to rotate. Because of the way the two electromagnets are connected, they are out of phase with one another by 90° and produce a rotating magnetic field. These electromagnets induce currents into the aluminum disk, which produces a magnetic field. This magnetic field interacts with the electromagnets' field, causing the disk to rotate. The number of rotations the motor makes is proportional to the current, voltage, and the time the motor is running (recall $E = I \times V \times t$). A counting mechanism records this number in terms of kilowatt-hours (kWh).

Fig. 15.26   An electric meter or an energy meter.

Measuring Temperature

Fig. 11.5   A thermocouple consists of two dissimilar pieces of wire connected to each other, producing a voltage proportional to the temperature. A reference junction scales the output to a known reference temperature.

Fig. 11.6   A thermistor is a resistive semiconductor device that is very sensitive to temperature. As the temperature changes, so will its resistance.

Fig. 11.7   A thermostat is a bimetal strip that uses the thermal expansion properties of metal as a temperature switch.

Optical pyrometers determine temperature by measuring the light emitted by a hot object. For instance, as a metal is heated, it changes color from red to white. Each shade of color corresponds to a particular temperature. The pyrometer converts the color it sees into a temperature.

Fig. 11.8   An optical pyrometer determines temperature by comparing the color of an internal filament to the color of the radiation emitted by an object whose temperature is being measured.

Measuring Volumetric Fluid Flow

To measure the volume of a fluid, a volumetric meter is used. A volumetric meter measures the volume of fluid flowing through it.

Fig. 10.7   As the water enters the volumetric flow meter, it fills chambers that have a known volume. As the chambers fill, they rotate, emptying their contents into the output. A recording mechanism records how many revolutions the chambers made.

Another type of volumetric flow meter commonly used in metering household water consumption uses a rotating impeller.

Fig. 10.8   Impeller-type volumetric fluid meter. As the fluid flows through the meter, the impeller rotates. The amount of rotations is proportional to the amount of fluid passing through it. A counting mechanism is mounted on the rotating shaft indicating the volume passed.

Measuring Fluid Flow

Fig. 10.9   A tachometer attached to an impeller-type flow meter indicates the rate of volume of fluid flowing through the meter.

Fig. 10.10   A thermal flow sensor contains a heater and an electronic thermometer placed next to one another. As the fluid flows over the heater, the heater cools. The greater the flow, the cooler the heater will be. The temperature is therefore a measure of the flow rate.

Measuring Light

Fig. 14.39   Light level can be measured using a photoresistor. A photoresistor has a light-sensitive material that changes resistance with light intensity.

Fig. 14.41   Typical photodiodes, which essentially are solar cells in a small package. They are used to detect light by putting out a voltage proportional to the light level.

Fig. 14.42   A phototransistor. Phototransistors have a light-sensitive base. Instead of a wire connected to the base, a window is present, allowing light through to turn on the transistor. Phototransistors have a larger output than a photodiode due to the amplifying nature of a transistor.

Measuring Sound

Fig. 13.37   A decibel meter.

Fig. 13.38   A crystal microphone. As the diaphragm moves back and forth from a sound wave, it causes the crystal to be flexed. Deforming this piezoelectric crystal produces a voltage across it proportional to the level of the sound wave.

Fig 13.39   A dynamic or moving coil microphone. As the diaphragm moves back and forth from a sound wave, the attached coil moves through a magnetic field, inducing a voltage into it.

Fig. 13.40   A condenser microphone measures sound levels using a capacitor. The diaphragm acts as a movable plate on a capacitor that responds to sound pressure.

Fig. 13.41   An electret-condenser microphone is essentially a condenser microphone with a permanent charge placed on the diaphragm at manufacturing. A built-in preamplifier, such as an FET, amplifies the small signals.

Density of a Fluid

Fig. 9.8   A hydrometer measures the specific gravity of a substance.

Measuring Magnetic Fields

Fig. 15.39   A hall probe responds to a magnetic field by generating a voltage proportional to the field strength. It is the most common method of measuring magnetic fields, and consists of a semiconductor material with a current running through it. If a magnetic field penetrates the probe, the current flowing inside the probe will be pushed to one side. This will produce a voltage across the probe because the charge will be concentrated on one side. This voltage is proportional to the magnetic field strength.

Fig. 15.16   A changing magnetic field will generate a current in the coil. Electromagnetic induction is the generation of electricity from magnetism. Magnetic fields can push electrons through a conductor just like a battery can. In order for the electricity to be generated, however, the magnetic field must be changing in some way, near the wire. Rotating a coil or wire near the magnet, or vice versa, will accomplish this.

Tables

Table C.1
English Weights and Measures

Units of length
Standard unit—inch (in. or ″)

$$12 \text{ inches} = 1 \text{ foot (f or ')}$$
$$3 \text{ feet} = 1 \text{ yard (yd)}$$
$$5\tfrac{1}{2} \text{ yards or } 16\tfrac{1}{2} \text{ feet} = 1 \text{ rod (rd)}$$
$$5280 \text{ feet} = 1 \text{ mile (mi)}$$

Units of weight
Standard unit—pound (Lb)

$$16 \text{ ounces (oz)} = 1 \text{ pound}$$
$$2000 \text{ pounds} = 1 \text{ ton (T)}$$

Units of volume
Liquid

$$16 \text{ ounces (fl oz)} = 1 \text{ pint (pt)}$$
$$2 \text{ pints} = 1 \text{ quart (qt)}$$
$$4 \text{ quarts} = 1 \text{ gallon (gal)}$$

Dry

$$2 \text{ pints (pt)} = 1 \text{ quart (qt)}$$
$$8 \text{ quarts} = 1 \text{ peck (pk)}$$
$$4 \text{ pecks} = 1 \text{ bushel (bu)}$$

Table C.2
Conversion Table for Length

	cm	m	km	in.	ft	mile
1 centimeter =	1	10^{-2}	10^{-5}	0.394	3.28×10^{-2}	6.21×10^{-6}
1 meter =	100	1	10^{-3}	39.4	3.28	6.21×10^{-4}
1 kilometer =	10^5	1000	1	3.94×10^4	3280	0.621
1 inch =	2.54	2.54×10^{-2}	2.54×10^{-5}	1	8.33×10^{-2}	1.58×10^{-5}
1 foot =	30.5	0.305	3.05×10^{-4}	12	1	1.89×10^{-4}
1 mile =	1.61×10^5	1610	1.61	6.34×10^4	5280	1

Table C.3
Conversion Table for Area

Metric	English
$1\ m^2 = 10{,}000\ cm^2$	$1\ ft^2 = 144\ in^2$
$= 1{,}000{,}000\ mm^2$	$1\ yd^2 = 9\ ft^2$
$1\ cm^2 = 100\ mm^2$	$1\ rd^2 = 30.25\ yd^2$
$= 0.0001\ m^2$	$1\ acre = 160\ rd^2$
$1\ km^2 = 1{,}000{,}000\ m^2$	$= 4840\ yd^2$
	$= 43{,}560\ ft^2$
	$1\ mi^2 = 640\ acres$

	m^2	cm^2	ft^2	in^2
1 square meter =	1	10^4	10.8	1550
1 square centimeter =	10^{-4}	1	1.08×10^{-3}	0.155
1 square foot =	9.29×10^{-2}	929	1	144
1 square inch =	6.45×10^{-4}	6.45	6.94×10^{-3}	1

1 circular mil = $5.07 \times 10^{-6}\ cm^2 = 7.85 \times 10^{-7}\ in^2$
1 hectare = $10{,}000\ m^2 = 2.47$ acres

Table C.4
Conversion Table for Volume

Metric	English
$1\ m^3 = 10^6\ cm^3$	$1\ ft^3 = 1728\ in^3$
$1\ cm^3 = 10^{-6}\ m^3$	$1\ yd^3 = 27\ ft^3$
$= 10^3\ mm^3$	

	m^3	cm^3	L	ft^3	in^3
$1\ m^3$	1	10^6	1000	35.3	6.10×10^4
$1\ cm^3 =$	10^{-6}	1	1.00×10^{-3}	3.53×10^{-5}	6.10×10^{-2}
1 litre =	1.00×10^{-3}	1000	1	3.53×10^{-2}	61.0
$1\ ft^3 =$	2.83×10^{-2}	2.83×10^4	28.3	1	1728
$1\ in^3 =$	1.64×10^{-5}	16.4	1.64×10^{-2}	5.79×10^{-4}	1

1 U.S. fluid gallon = 4 U.S. fluid quarts = 8 U.S. pints = 128 U.S. fluid ounces = $231\ in^3 = 0.134\ ft^3$
1 L = $1000\ cm^3 = 1.06$ qt   1 fl oz = $29.5\ cm^3$
$1\ ft^3 = 7.47$ gal $= 28.3$ L

Table C.5
Conversion Table for Mass

	g	kg	slug	oz	Lb	ton
1 gram =	1	0.001	6.85×10^{-5}	3.53×10^{-2}	2.21×10^{-3}	1.10×10^{-6}
1 kilogram =	1000	1	6.85×10^{-2}	35.3	2.21	1.10×10^{-3}
1 slug =	1.46×10^4	14.6	1	515	32.2	1.61×10^{-2}
1 ounce =	28.4	2.84×10^{-2}	1.94×10^{-3}	1	6.25×10^{-2}	3.13×10^{-5}
1 pound =	454	0.454	3.11×10^{-2}	16	1	5.00×10^{-4}
1 ton =	9.07×10^5	907	62.2	3.2×10^4	2000	1

1 metric ton = 1000 kg = 2205 Lb

Quantities in the shaded areas are not mass units. When we write, for example, 1 kg "=" 2.21 lb, this means that a kilogram is a mass that weighs 2.21 pounds under standard conditions of gravity (g = 9.80 m/s^2 = 32.2 ft/s^2).

Table C.6
Conversion Table for Density

	slug/ft^3	kg/m^3	g/cm^3	Lb/ft^3	Lb/in^3
1 slug per ft^3 =	1	515	0.515	32.2	1.86×10^{-2}
1 kilogram per m^3 =	1.94×10^{-3}	1	0.001	6.24×10^{-2}	3.61×10^{-5}
1 gram per cm^3 =	1.94	1000	1	62.4	3.61×10^{-2}
1 pound per ft^3 =	3.11×10^{-2}	16.0	1.60×10^{-2}	1	5.79×10^{-4}
1 pound per in^3 =	53.7	2.77×10^4	27.7	1728	1

Quantities in the shaded areas are weight densities and, as such, are dimensionally different from mass densities.

Note that $D_w = D_m g$ where

$$D_w = \text{weight density}$$

$$D_m = \text{mass density}$$

$$g = 9.80 \text{ m/s}^2 = 32.2 \text{ ft/s}^2$$

Table C.7
Conversion Table for Time

	yr	day	h	min	s
1 year =	1	365	8.77×10^3	5.26×10^5	3.16×10^7
1 day =	2.74×10^{-3}	1	24	1440	8.64×10^4
1 hour =	1.14×10^{-4}	4.17×10^{-2}	1	60	3600
1 minute =	1.90×10^{-6}	6.94×10^{-4}	1.67×10^{-2}	1	60
1 second =	3.17×10^{-8}	1.16×10^{-5}	2.78×10^{-4}	1.67×10^{-2}	1

Table C.8
Conversion Table for Speed

	ft/s	km/h	m/s	mi/h	cm/s
1 foot per second =	1	1.10	0.305	0.682	30.5
1 kilometer per hour =	0.911	1	0.278	0.621	27.8
1 meter per second =	3.28	3.60	1	2.24	100
1 mile per hour =	1.47	1.61	0.447	1	44.7
1 centimeter per second =	3.28×10^{-2}	3.60×10^{-2}	0.01	2.24×10^{-2}	1

1 mi/min = 88.0 ft/s = 60.0 mi/h

Table C.9

Conversion Table for Force

	N	Lb	oz
1 newton =	1	0.225	3.60
1 pound =	4.45	1	16
1 ounce =	0.278	0.0625	1

Table C.10

Conversion Table for Power

	Btu/h	ft lb/s	hp	kW	W
1 British thermal unit per hour =	1	0.216	3.93×10^{-4}	2.93×10^{-4}	0.293
1 foot pound per second =	4.63	1	1.82×10^{-3}	1.36×10^{-3}	1.36
1 horsepower =	2550	550	1	0.746	746
1 kilowatt =	3410	738	1.34	1	1000
1 watt =	3.41	0.738	1.34×10^{-3}	0.001	1

Table C.11

Conversion Table for Pressure

	atm	Inch of Water	mm Hg	N/m^2 (Pa)	Lb/in^2	Lb/ft^2
1 atmosphere =	1	407	760	1.01×10^5	14.7	2120
1 inch of water[a] at 4°C =	2.46×10^{-3}	1	1.87	249	3.61×10^{-2}	5.20
1 millimeter of mercury[a] at 0°C =	1.32×10^{-3}	0.535	1	133	1.93×10^{-2}	2.79
1 newton per meter^2 (pascal) =	9.87×10^{-6}	4.02×10^{-3}	7.50×10^{-3}	1	1.45×10^{-4}	2.09×10^{-2}
1 pound per in^2 =	6.81×10^{-2}	27.7	51.7	6.90×10^3	1	144
1 pound per ft^2 =	4.73×10^{-4}	0.192	0.359	47.9	6.94×10^{-3}	1

[a]Where the acceleration of gravity has the standard value, $9.80 \text{ m/s}^2 = 32.2 \text{ ft/s}^2$.

Table C.12
Mass and Weight Density[a]

Substance	Mass Density (kg/m³)	Weight Density (Lb/ft³)
Solids		
Copper	8,890	555
Iron	7,800	490
Lead	11,300	708
Aluminum	2,700	169
Ice	917	57
Wood, white pine	420	26
Concrete	2,300	145
Cork	240	15
Liquids		
Water	$1,000^b$	62.4
Seawater	1,025	64.0
Oil	870	54.2
Mercury	13,600	846
Alcohol	790	49.4
Gasoline	680	42.0

	At 0°C and 1 atm Pressure	At 32°F and 1 atm Pressure
Gases[a]		
Air	1.29	0.081
Carbon dioxide	1.96	0.123
Carbon monoxide	1.25	0.078
Helium	0.178	0.011
Hydrogen	0.0899	0.0056
Oxygen	1.43	0.089
Nitrogen	1.25	0.078
Ammonia	0.760	0.047
Propane	2.02	0.126

[a]The density of a gas is found by pumping the gas into a container, by measuring its volume and mass or weight, and then by using the appropriate density formula.

[b]Metric weight density of water $= 9800$ N/m³.

Table C.13
Specific Gravity of Certain Liquids at
Room Temperature (20°C or 68°F)

Liquid	Specific Gravity
Benzene	0.90
Ethyl alcohol	0.79
Gasoline	0.68
Kerosene	0.82
Mercury	13.6
Seawater	1.025
Sulfuric acid	1.84
Turpentine	0.87
Water	1.000

Table C.14
Conversion Table for Energy, Work, and Heat

	Btu	ft Lb	J	cal	kWh
1 British thermal unit =	1	778	1060	252	2.93×10^{-4}
1 foot pound =	1.29×10^{-3}	1	1.36	0.324	3.77×10^{-7}
1 joule =	9.48×10^{-4}	0.738	1	0.239	2.78×10^{-7}
1 calorie =	3.97×10^{-3}	3.09	4.19	1	1.16×10^{-6}
1 kilowatt-hour =	3410	2.66×10^{6}	3.60×10^{6}	8.60×10^{5}	1

Table C.15
Heat Constants

	Melting Point (°C)	Boiling Point (°C)	Specific Heat cal/g °C or kcal/kg °C or Btu/Lb °F	J/kg °C	Heat of Fusion cal/g or kcal/kg	J/kg	Heat of Vaporization cal/g or kcal/kg	J/kg
Alcohol, ethyl	−117	78.5	0.58	2400	24.9	1.04×10^{5}	204	8.54×10^{5}
Aluminum	660	2057	0.22	920	76.8	3.21×10^{5}		
Brass	840		0.092	390				
Copper	1083	2330	0.092	390	49.0	2.05×10^{5}		
Glass			0.21	880				
Ice	0		0.51	2100	$8\overline{0}$	3.35×10^{5}		
Iron (steel)	1540	3000	0.115	481	7.89	3.30×10^{4}		
Lead	327	1620	0.031	130	5.86	2.45×10^{4}		
Mercury	−38.9	357	0.033	140	2.82	1.18×10^{4}	65.0	2.72×10^{5}
Silver	961	1950	0.056	230	26.0	1.09×10^{5}		
Steam			0.48	2000				
Water (liquid)	0	$10\overline{0}$	1.00	4190	$8\overline{0}$	3.35×10^{5}	$54\overline{0}$	2.26×10^{6}
Zinc	419	907	0.092	390	23.0	9.63×10^{4}		

Table C.16
Coefficient of Linear Expansion

Material	α (metric)	α (English)
Aluminum	$2.3 \times 10^{-5}/°C$	$1.3 \times 10^{-5}/°F$
Brass	$1.9 \times 10^{-5}/°C$	$1.0 \times 10^{-5}/°F$
Concrete	$1.1 \times 10^{-5}/°C$	$6.0 \times 10^{-6}/°F$
Copper	$1.7 \times 10^{-5}/°C$	$9.5 \times 10^{-6}/°F$
Glass	$9.0 \times 10^{-6}/°C$	$5.1 \times 10^{-6}/°F$
Pyrex	$3.0 \times 10^{-6}/°C$	$1.7 \times 10^{-6}/°F$
Steel	$1.3 \times 10^{-5}/°C$	$6.5 \times 10^{-6}/°F$
Zinc	$2.6 \times 10^{-5}/°C$	$1.5 \times 10^{-5}/°F$

Table C.17
Coefficient of Volume Expansion

Liquid	β (metric)	β (English)
Acetone	$1.49 \times 10^{-3}/C°$	$8.28 \times 10^{-4}/F°$
Alcohol, ethyl	$1.12 \times 10^{-3}/C°$	$6.62 \times 10^{-4}/F°$
Carbon tetrachloride	$1.24 \times 10^{-3}/C°$	$6.89 \times 10^{-4}/F°$
Mercury	$1.8 \times 10^{-4}/C°$	$1.0 \times 10^{-4}/F°$
Petroleum	$9.6 \times 10^{-4}/C°$	$5.33 \times 10^{-4}/F°$
Turpentine	$9.7 \times 10^{-4}/C°$	$5.39 \times 10^{-4}/F°$
Water	$2.1 \times 10^{-4}/C°$	$1.17 \times 10^{-4}/F°$

Table C.18
Conversion Table for Charge

Charge on one electron $= 1.60 \times 10^{-19}$ coulomb

1 coulomb $= 6.25 \times 10^{18}$ electrons of charge

1 ampere-hour $= 3600$ C

Table C.19
Copper Wire Table

| Gauge No. | Diameter (mils) | Diameter (mm) | Cross Section | | Ohms per 1000 ft | | Weight per 1000 ft (Lb) |
			cir mils	in²	25°C (77°F)	65°C (149°F)	
0000	460.0		212,000	0.166	0.0500	0.0577	641.0
000	410.0		168,000	0.132	0.0630	0.0727	508.0
00	365.0		133,000	0.105	0.0795	0.0917	403.0
0	325.0		106,000	0.0829	0.100	0.116	319.0
1	289.0	7.35	83,700	0.0657	0.126	0.146	253.0
2	258.0	6.54	66,400	0.0521	0.159	0.184	201.0
3	229.0	5.83	52,600	0.0413	0.201	0.232	159.0
4	204.0	5.19	41,700	0.0328	0.253	0.292	126.0
5	182.0	4.62	33,100	0.0260	0.319	0.369	100.0
6	162.0	4.12	26,300	0.0206	0.403	0.465	79.5
7	144.0	3.67	20,800	0.0164	0.508	0.586	63.0
8	128.0	3.26	16,500	0.0130	0.641	0.739	50.0
9	114.0	2.91	13,100	0.0103	0.808	0.932	39.6
10	102.0	2.59	10,400	0.00815	1.02	1.18	31.4
11	91.0	2.31	8,230	0.00647	1.28	1.48	24.9
12	81.0	2.05	6,530	0.00513	1.62	1.87	19.8
13	72.0	1.83	5,180	0.00407	2.04	2.36	15.7
14	64.0	1.63	4,110	0.00323	2.58	2.97	12.4
15	57.0	1.45	3,260	0.00256	3.25	3.75	9.86
16	51.0	1.29	2,580	0.00203	4.09	4.73	7.82
17	45.0	1.15	2,050	0.00161	5.16	5.96	6.20
18	40.0	1.02	1,620	0.00128	6.51	7.51	4.92
19	36.0	0.91	1,290	0.00101	8.21	9.48	3.90
20	32.0	0.81	1,020	0.000802	10.4	11.9	3.09
21	28.5	0.72	810	0.000636	13.1	15.1	2.45
22	25.3	0.64	642	0.000505	16.5	19.0	1.94
23	22.6	0.57	509	0.000400	20.8	24.0	1.54
24	20.1	0.51	404	0.000317	26.2	30.2	1.22
25	17.9	0.46	320	0.000252	33.0	38.1	0.970
26	15.9	0.41	254	0.000200	41.6	48.0	0.769
27	14.2	0.36	202	0.000158	52.5	60.6	0.610
28	12.6	0.32	160	0.000126	66.2	76.4	0.484
29	11.3	0.29	127	0.0000995	83.4	96.3	0.384
30	10.0	0.26	101	0.0000789	105	121	0.304
31	8.9	0.23	79.7	0.0000626	133	153	0.241
32	8.0	0.20	63.2	0.0000496	167	193	0.191
33	7.1	0.18	50.1	0.0000394	211	243	0.152
34	6.3	0.16	39.8	0.0000312	266	307	0.120
35	5.6	0.14	31.5	0.0000248	335	387	0.0954
36	5.0	0.13	25.0	0.0000196	423	488	0.0757
37	4.5	0.11	19.8	0.0000156	533	616	0.0600
38	4.0	0.10	15.7	0.0000123	673	776	0.0476
39	3.5	0.09	12.5	0.0000098	848	979	0.0377
40	3.1	0.08	9.9	0.0000078	1070	1230	0.0200

Table C.20
Conversion Table for Plane Angles

	°	′	″	rad	rev
1 degree =	1	60	3600	1.75×10^{-2}	2.78×10^{-3}
1 minute =	1.67×10^{-2}	1	60	2.91×10^{-4}	4.63×10^{-5}
1 second =	2.78×10^{-4}	1.67×10^{-2}	1	4.85×10^{-6}	7.72×10^{-7}
1 radian =	57.3	3440	2.06×10^5	1	0.159
1 revolution =	360	2.16×10^4	1.30×10^6	6.28 or 2π	1

Table C.21
The Greek Alphabet

Capital	Lowercase	Name
A	α	alpha
B	β	beta
Γ	γ	gamma
Δ	δ	delta
E	ϵ	epsilon
Z	ζ	zeta
H	η	eta
Θ	θ	theta
I	ι	iota
K	κ	kappa
Λ	λ	lambda
M	μ	mu
N	ν	nu
Ξ	ξ	xi
O	o	omicron
Π	π	pi
P	ρ	rho
Σ	σ	sigma
T	τ	tau
Υ	υ	upsilon
Φ	ϕ	phi
X	χ	chi
Ψ	ψ	psi
Ω	ω	omega

Table C.22

Characteristics of Some Common Batteries

Battery	Voltage	Ampere-hour rating	Mass, g
Primary			
"pen-light"	1.5	0.58	20
D-cell	1.5	3.0	90
#6 dry cell	1.5	30	860
#2U6 battery	9.0	0.325	30
#V-60 battery	90	0.47	580
#1 mercury cell	1.4	1.0	12
#42 mercury cell	1.4	14	166
Secondary (rechargeable)			
#CD-21 nickel-cadmium	6.0	0.15	52
#CD-29 nickel-cadmium	12.0	0.45	340
lead-acid car battery	6.0	84	13 000
lead-acid car battery	12.0	96	25 000

Table C.23

Coefficients of Static Friction and Sliding Friction

Surfaces in contact	Coefficient of static friction	
	Dry	Well lubricated
Hard steel and babbitt	0.42	0.17
Hard steel and hard steel	0.78	0.11
Soft steel and soft steel	0.74	0.11
Soft steel and lead	0.95	0.5
Plastic and steel	0.5	0.1
Rubber and concrete	1.5	1.0 (wet)

Surfaces in contact	Coefficient of sliding friction		
	Dry	Lightly lubricated	Well lubricated
Bronze and cast iron	0.21	0.16	0.077
Cast iron and cast iron	0.4	0.15	0.064
Cast iron and hardwood	0.49	0.19	0.075
Hardwood and hardwood	0.48	0.16	0.067
Hard steel and babbitt	0.35	0.16	0.06
Hard steel and hard steel	0.42	0.12	0.03
Plastic and steel	0.35	—	0.05
Rubber and concrete	1.02	0.9 (wet)	—
Soft steel and soft steel	0.57	0.19	0.09

Values listed are for ordinary pressures and temperatures, but may still vary somewhat from sample to sample.

Table C.24
Elastic Limit and Ultimate Strength of Some Common Solids

Solid	Elastic limit		Ultimate strength	
	Tension	Compression	Tension	Compression
Aluminum	840	840	1 800	∞
Brass	630	630	2 000	∞
Brick, best hard	30	840	30	840
Brick, common	4	70	4	70
Bronze	2 800	2 800	5 300	∞
Cement, Portland, 1 mo old	28	140	28	140
Cement, Portland, 1 yr old	35	210	35	210
Concrete, Portland, 1 mo old	14	70	14	70
Concrete, Portland, 1 yr old	28	140	28	140
Copper	700	700	2 500	∞
Douglas fir	330	330	500	430
Granite	49	1 300	49	1300
Iron, cast	420	1 800	1 400	5600
Lead	10	10	200	∞
Limestone and sandstone	21	630	21	630
Monel metal	6 300	6 300	7 000	∞
Oak, white	310	310	600	520
Pine, white, eastern	270	270	400	345
Slate	35	980	35	980
Steel, bridge cable	6 700	6 700	15 000	∞
Steel, 1 percent C, tempered	5 000	5 000	8 400	8 400
Steel, chrome, tempered	9 100	9 100	11 000	11 000
Steel, stainless	2 100	2 100	5 300	5 300
Steel, structural	2 500	2 500	4 600	4 600

Units are kilogram-force per square centimeter (kg_f/cm^2).

Note: $1 \dfrac{kg_f}{cm^2} = 98.1 \text{ kPa} = 14.22 \dfrac{Lb}{in^2}$

Table C.25
Hardness of Materials

Aluminum	2–2.9
Brass	3–4
Copper	2.5–3
Diamond	10
Gold	2.5–3
Iron	4–5
Lead	1.5
Magnesium	2.0
Marble	3–4
Mica	2.8
Phosphorbronze	4
Platinum	4.3
Quartz	7
Silver	2.5–4
Steel	5–8.5
Tin	1.5–1.8
Zinc	2.5

Table C.26
Elastic Constants for Various Materials in SI Units and USCS Units

Material	Young's modulus Y, MPa*	Shear modulus S, MPa	Elastic limit, MPa	Ultimate strength, MPa
Aluminum	68,900	23,700	131	145
Brass	89,600	35,300	379	455
Copper	117,000	42,300	159	338
Iron	89,600	68,900	165	324
Steel	207,000	82,700	248	489

*1 MPa = 10^6 Pa.

Material	Young's modulus Y, Lb/in.^{2}	Shear modulus S, Lb/in.^{2}	Elastic limit, Lb/in.^{2}	Ultimate strength, Lb/in.^{2}
Aluminum	10×10^6	3.44×10^6	19,000	21,000
Brass	13×10^6	5.12×10^6	55,000	66,000
Copper	17×10^6	6.14×10^6	23,000	49,000
Iron	13×10^6	10×10^6	24,000	47,000
Steel	30×10^6	12×10^6	36,000	71,000

Table C.27

R-Values for Some Common Building Materials

Material	R-value (ft² · °F · h/Btu)
Hardwood siding (1.0 in thick)	0.91
Wood shingles (lapped)	0.87
Brick (4.0 in thick)	4.00
Concrete block (filled cores)	1.93
Styrofoam (1.0 in thick)	5.0
Fiberglass batting (3.5 in thick)	10.90
Fiberglass batting (6.0 in thick)	18.80
Fiberglass board (1.0 in thick)	4.35
Cellulose fiber (1.0 in thick)	3.70
Flat glass (0.125 in thick)	0.89
Insulating glass (0.25-in space)	1.54
Vertical air space (3.5 in thick)	1.01
Air film	0.17
Dry wall (0.50 in thick)	0.45
Sheathing (0.50 in thick)	1.32

Table C.28

Typical Heats of Combustion of Some Gaseous Fuels*

Fuel	Heat of combustion				Air-fuel ratio for complete combustion, ft³/ft³ or L/L
	On mass basis		On volume basis		
	Btu/Lb$_m$	MJ/kg	Btu/ft³	kJ/L	
Hydrogen	61 400	143.0	275	10.3	2.4
Methane	23 800	55.2	900	33	9.6
Propane	22 200	51.5	2400	88	13.7
Natural gas	23 600	54.8	1000	37	11.9
Coke gas	23 000	53.6	1300	50	14

*Data applies to gases at 1.0 atm pressure and 15.6°C [60°F]. Actual values vary slightly depending on geographical origin. Values do not include latent heat in any water vapor formed as a product of combustion.

Note: 1 MJ/kg = 1000 kJ/kg = 1000 J/g

      1 kJ/L = 1 J/cm³ = 1 MJ/m³

Table C.29
Typical Heats of Combustion of Some Solid and Liquid Fuels*

Fuel	Heat of combustion		Air-fuel ratio for complete combustion, Lb_m/Lb_m or g/g
	Btu/Lb_m	MJ/kg	
Butane	20 000	46	15.4
Gasoline	19 000	44	14.8
Crude oil	18 000	42	14.2
Fuel oil	17 500	41	13.8
Coke	14 500	34	11.3
Coal, bituminous	13 500	31	10.4

*Actual values will vary slightly depending on geographical origin. Values do not include latent heat in any water vapor formed as a product of combustion.

Note: 1 MJ/kg = 1000 kJ/kg = 1000 J/g

Table C.30
Speed of Sound in Various Media

Medium	Speed	
	m/s	ft/s
Aluminum	6,420	21,100
Brass	4,700	15,400
Steel	5,960	19,500
Granite	6,000	19,700
Alcohol	1,210	3,970
Water (25°C)	1,500	4,920
Air, dry (0°C)	331	1,090
Vacuum	0	0

Table C.31
Indices of Refraction*

Material	$n = \dfrac{c}{v}$
Vacuum	1.0000
Air (at STP)	1.0003
Water	1.33
Ethyl alcohol	1.36
Glass	
Fused quartz	1.46
Crown glass	1.52
Light flint	1.58
Lucite or Plexiglas	1.51
Sodium chloride	1.53
Diamond	2.42

*$\lambda = 589$ nm.

Labs

Contained in this appendix are ten laboratory experiments.

 ## MEASURING LENGTH

Objective: To build and calibrate a device to measure length using a wire's resistance dependence on length.

Theory: A wire's resistance changes with its length. By measuring the resistance of a wire of a known length, we can determine the wire's resistivity ρ (the number of ohms per meter of wire). The resistance is given by:

$$R = \rho \times L$$

$$\text{Resistance } (\Omega) = \text{resistivity} \left(\frac{\Omega}{m}\right) \times \text{length (m)}$$

Once we know the wire's resistivity, we can use the wire to measure the length of an object. By measuring the resistance of the wire with the same length as the object, we can determine its length from the equation:

$$L = \frac{R}{\rho}$$

Procedure:

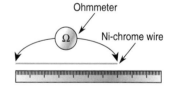

Fig. D1

1. Determine the wire's resistivity by measuring the wire's resistance for known lengths. (For best results, use a fine gauge wire made of ni-chrome.)

L (m)	R (Ω)	$\rho\left(\dfrac{\Omega}{m}\right) = \dfrac{R}{L}$
1 m		
0.9 m		
0.8 m		
0.7 m		

From these four measurements calculate the average value of ρ.

$$\rho_{average} = \underline{\quad\quad}$$

2. Check the accuracy of your wire length measurement device by measuring the length of some objects using the wire and compare your results with their real length using a meter stick.

Object #	$R\,(\Omega)$	$L = \dfrac{R}{\rho_{average}}$	Length measured with a meter stick	% Difference = $\left\|\dfrac{L_{meterstick} - L_{resistance}}{L_{meterstick}}\right\| \times 100$
1				
2				
3				
4				

Objective: To make a clock using a pendulum and determine which factors influence its period of oscillation.

Theory: A pendulum's period of oscillation depends only on its length and gravity and can be approximated for small oscillations by:

$$T = 2\pi\sqrt{\frac{L}{g}}$$

$$\text{Period of one oscillation} = 2\pi\sqrt{\frac{\text{length}}{\text{gravity}}}$$

An *oscillation* is one complete back-and-forth motion.

Procedure:

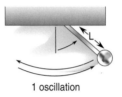

1 oscillation

Fig. D2

1. Show that the period does not depend on the mass. Explain how you did this.

2. If $L = 1$ m what is T?

3. Show by experiment the answer you calculated in step 2 is correct. Explain what you did.

4. Use your pendulum as a clock to determine your heart rate in beats per minute. Explain how you did this. What is your rate?

MEASURING g

Objective: To measure free-fall acceleration.

Theory: Everything falls at the same rate (9.8 m/s^2) independent of its mass (neglecting air friction).

The equation of motion:

$$d = v_i \cdot t + \frac{1}{2} g \cdot t^2$$

can be used to determine free-fall acceleration (g). If $v_i = 0$, then:

$$g = \frac{2d}{t^2}$$

Procedure:

1. Verify that different masses fall at the same rate. Explain how you did this.

2. Time how long it takes a mass to fall from rest from a measured distance. Do this for several trials and calculate g and the percent discrepancy from the measured and the actual $g = 9.8$ m/s^2.

d (meters)	t (seconds)	Measured g $g = 2d/t^2$	% difference $=$ $\left\|\dfrac{g_{measured} - 9.8}{9.8}\right\| \times 100$

3. How could you measure g more accurately?

4. Are there any real applications for measuring variations in g over the surface of the earth?

MEASURING FORCE

Objective: To build and calibrate a force sensor using conductive foam.

Theory: A simple force sensor can be constructed out of conductive foam (integrated circuit chips are stored on this foam to avoid damage done by static charge). As the foam is compressed, its resistance decreases. Using this property, we can calibrate this force sensor, developing a relationship between the force on the sensor and its resistance.

Procedure:

Fig. D3

1. Record the resistance of the foam as weights are placed on top of the foam.

Weight	Resistance

2. Plot your data, resistance versus weight.
3. Place some unknown weights on your sensor. From your graph determine the value of these weights and compare this value with the real value.

Weight #	Resistance	Weight determined from graph	Real value of weight	% Difference = $\left\| \dfrac{weight_{graph} - weight_{real\ value}}{weight_{real\ value}} \right\| \times 100$
1				
2				
3				
4				

4. What are the problems with this simple sensor, and how could you alleviate them?

Objective: To measure the horsepower of a small motor using a homemade dynamometer.

Theory: By applying a force to the spinning shaft of a motor and measuring its angular speed, we can determine its horsepower.

Power is given by the formula:

$$P = F \cdot v$$

Power (ft Lb/s) = force (Lb) × velocity (ft/s)

The tangential velocity of a spinning shaft is given by:

$$v = 2\pi r \cdot f$$

Velocity = 2π × radius (ft) × frequency (revolutions/second)

Therefore:

$$P = F \cdot 2\pi r \cdot f$$

To convert power in ft Lb/s into horsepower use:

$$Hp = P(\text{ft Lb/s}) \times \left(\frac{1 \text{ hp}}{550 \text{ ft Lb/s}} \right)$$

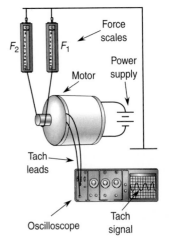

Fig. D4

Procedure:

1. Using a motor with a built-in tachometer or a means to measure angular speed, set up the following:

2. Arrange the motor so that it sits on the tabletop with the string looped under the motor's shaft with tension in the spring scales.

3. Apply a voltage to the motor and measure F_1, F_2, and the motor's revolutions per second.

4. Record these measurements for different applied voltages. Generate the following table.

Voltage (v)	F_1 (Lb)	F_2 (Lb)	f = Rev/sec	$P = \|F_2 - F_1\| \cdot$ $2\pi r \cdot f \left(\frac{\text{ft Lb}}{\text{s}} \right)$	P (Hp)

5. Electrical power in watts is given by:

$$P = I \cdot V$$

Power (watts) = current × voltage

Calculate the electrical power consumed by the motor by measuring the current for an applied voltage. Using the conversion 1 hp = 746 W, determine the horsepower.

$$\text{Hp} = I \times V \times \left(\frac{1 \text{ hp}}{746 \text{ W}}\right) = \underline{\qquad}$$

6. Determine the efficiency of the motor.

$$\text{Efficiency} = \frac{\text{power}_{\text{out}}}{\text{power}_{\text{in}}} \times 100\%$$

$$\text{Efficiency} = \frac{\text{power measured by the dynamometer}}{\text{power electrical}} \times 100\% =$$

7. Is your motor very efficient? Why or why not?

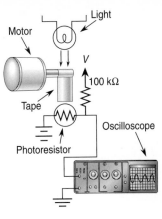

Fig. D5

Objective: To measure the angular velocity of a rotating shaft.

Theory: A tachometer is a device that measures angular velocity (RPM). A simple tachometer can be made from a photoresistor, a light source, and monitored with an oscilloscope. Fig. D5 shows a rotating shaft with tape attached to it.

Once every revolution, the light source is blocked from the photo resistor by the tape. The photoresistor is used in a voltage divider circuit and monitored with a scope. The period of the waveform on the scope can be used to determine the angular velocity of the shaft.

$$\omega = \frac{1 \text{ rev}}{T}$$

$$\text{Angular velocity (rev/s)} = \frac{1 \text{ revolution}}{\text{period of one revolution (second)}}$$

To determine RPM use:

$$\text{RPM} = \frac{1 \text{ rev}}{T}\left(\frac{60 \text{ s}}{1 \text{ min}}\right)$$

Procedure:

1. Build the setup pictured above and measure the angular velocity of the motor.
2. Determine how the motor's RPM changes with applied voltage to the motor.

Voltage (V)	T (seconds)	$\text{RPM} = \dfrac{1}{T} \times 60$

3. Plot RPM versus applied voltage.
4. How else could you measure the RPM of this motor?

Objective: To demonstrate several simple machines and measure their mechanical advantage.

Theory: The mechanical advantage of a machine is a measure of how much force it can generate compared with the input force into the machine. It is given by:

$$MA = \frac{F_{out}}{F_{in}}$$

The mechanical advantage can also be expressed by:

$$MA = \text{efficiency} \times \frac{d_{in}}{d_{out}}$$

where d_{in} and d_{out} are the input and output distances of the machine and the efficiency is related to any friction involved in operating the machine.

We can put these two equations together to calculate the efficiency.

$$\text{Efficiency} = \left(\frac{F_{out}/F_{in}}{d_{in}/d_{out}}\right) \times 100$$

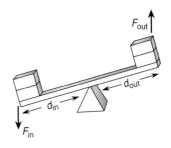

Fig. D6

Procedure:

Determine the mechanical advantage and the efficiency of the machines below.

1. The lever. Experiment with the lever for various lengths of d_{in} and d_{out}.

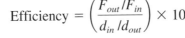

d_{in}	d_{out}	F_{in}	F_{out}	$MA = F_{out}/F_{in}$	Efficiency $= \left(\frac{F_{out}/F_{in}}{d_{in}/d_{out}}\right) \times 100\%$

2. The Pulley. Put together the following configurations and fill in the table below.

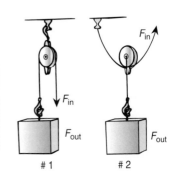

#1   #2

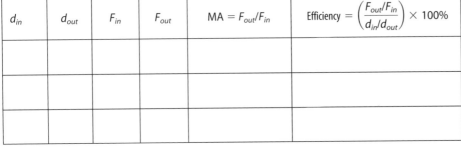

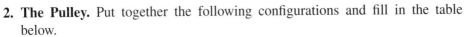

Configuration	d_{in}	d_{out}	F_{in}	F_{out}	$MA = F_{out}/F_{in}$	Efficiency $= \left(\frac{F_{out}/F_{in}}{d_{in}/d_{out}}\right) \times 100$
#1						
#2						
#3						
#4						

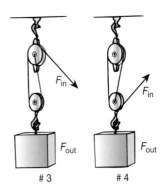

#3   #4

Fig. D7

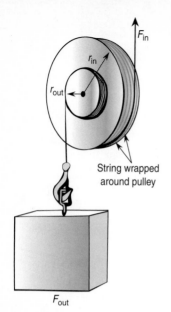

F_{in}

r_{in}

r_{out}

String wrapped
around pulley

F_{out}

Fig. D8

3. The wheel and axle. Experiment for various radii of r_{in} and r_{out}.

r_{in}	r_{out}	F_{in}	F_{out}	MA = F_{out}/F_{in}	Efficiency = $\left(\dfrac{F_{out}/F_{in}}{r_{in}/r_{out}}\right) \times 100$

MEASURING TEMPERATURE

Objective: To measure and calibrate temperature sensors using a thermistor and a thermocouple.

Theory: Temperature changes the electrical properties of materials. A thermistor is a semiconductor that changes its resistance with temperature. A thermocouple consists of two dissimilar pieces of wire twisted together at one end, which when heated produces a voltage across the wires.

Procedure:

Set up the following:

Calibrate the thermistor and thermocouple by placing them on a soldering iron along with a thermometer. Plug in the iron and while it is heating up, record the temperature of the thermometer, voltage of the thermocouple, and resistance of the thermistor. Fill in the table below.

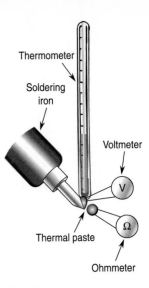

Fig. D9

Temperature	Resistance	Voltage

- Plot resistance versus temperature.
- Plot voltage versus temperature.

Use your sensors to measure the room's temperature and your temperature. Check with your thermometer that these are correct.

MEASURING LIGHT

Objective: To measure light intensity using a photoresistor, photodiode, and solar cell.

Theory: Some materials change their electrical properties with light. A photoresistor lowers its resistance with an increasing light level. Solar cells and LEDs generate a voltage with light.

Procedure: Set up the following circuits using an LED of a known light output (usually this is listed as so many mcd for a given applied voltage) as a source of light and measure how the sensors respond to this light.

Record how the sensors respond to various intensities of light.

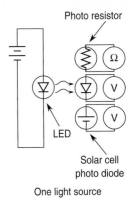

One light source

Number of light sources	Intensity	Resistance of photoresistor	Voltage, LED	Voltage, solar cell
1				
2				
3				
4				

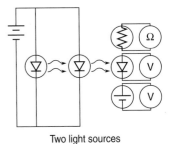

Two light sources

Fig. D10

1. Plot resistance versus intensity.

2. Plot voltage of the photodiode versus intensity.

3. Plot voltage of the solar cell versus intensity.

4. From your calibrated sensors, can you determine the intensity of the light in the room? What is this value?

DC MOTORS

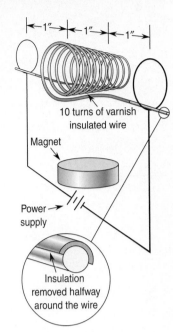

10 turns of varnish insulated wire

Magnet

Power → supply

Insulation removed halfway around the wire

Fig. D11

Objective: To build a simple DC motor.

Theory: Magnetic like poles repel, and unlike poles attract. An electric motor uses this principle to alternately repel and attract the magnetic field generated in a current-carrying coil to a stationary permanent magnet in order to produce a rotation.

Procedure:

1. Using a small-gauge varnished magnet wire such as 22-gauge or smaller, wind a small coil of approximately 10 turns and 1 inch in diameter. Taking the ends of the coil, wrap them around the coil to hold the turns together. With 1 inch of wire protruding from opposite ends of the coil, sand off the varnish from each end but only halfway around the wire.

2. Using two other pieces of wire, remove the varnish insulation and create a small loop on each end. These pieces will support the coil and deliver current to it. Attach or connect these two wires to the ends of a D-cell or power supply.

3. Place the coil through the support wires and place a magnet under the coil. Spin the coil with your finger. If made correctly, the coil should continue to spin on its own.

Questions:

1. Explain why the coil spins on its own.
2. How could you make this motor spin faster and smoother?

INDEX

Converging lenses, 212
Conversation, 198
Conversions, 19
   area, 26
   multiple, 20
   volume, 28
Conveyer belt, 51, 103
Coefficient of thermal expansion,
      181
Colors, 204
Cooler electronic, 180
Coordinate systems, 37, 52
Copper, 220
Corner reflector, 210
Cosine, 236
Cost, 10
Coulomb, 77
Counter balance, 59
Counter current, 5
Crankshaft, 64
CRT, 46, 47
Cryogenic, 222
Current, 3
Cycle, 187

D

Deceleration, 41
Decibel meter, 198
Decompressing, 194
Degree
   unit of angular measure, 91
   unit of temperature, 151
Density, 29
Depth, 135
Dielectric, 23
Diesel engine, 177
Diffraction, 210
Dispersion, 213, 214
Displacement, 97
Distance measurements along a
      circle, 93
Disturbance, 186
Diverging lenses, 212
Dog house, 170
Dominoes, 186
Door latch, 6
Doppler effect, 41
Doppler shift, 192
Doppler radar, 192
Driven
   gear, 97
   wheel, 97
Driving
   gear, 97
   wheel, 97
Ductility, 127
DVD, 215
Dynamic load, 60

E

Eardrum, 196
Efficiency, 73, 79, 106, 125,
      178
Elastic limit, 120
Electric
   field, 2, 3
   meter, 229
   power, 86
Electrical
   charges, 1
   discharge, 206
   energy, 10
   generator, 5, 87, 89, 225
   power, 10
   pressure, 3
   units, 76
Electricity, 1
Electromagnet, 221
Electromagnetic
   induction, 5, 225
   spectrum, 203
   waves, 203
   speed of, 206
   generation of, 204
Electron, 1, 2
Electron gun, 46
Elevator, 87
Energy
   conservation of, 81
   defined, 70
   electric, 10
   heat, 155
   kinetic, 79
   potential, 74
   sources, 74
   loss, 73
   transfer, 186
   waves in, 186
English system of units, 15
Evaporation, 168
Excited state, 205
Expansion, thermal
   area, 182
   coefficient of, 181
   linear, 181
   of solids, of volume, 182
External combustion engine, 177
Exponent, 14
Eye, 212

F

Fahrenheit scale, 151
Farad, 5
Ferric oxide, 229
Ferromagnetic, 230
Fetus, 196

Field
   electric, 2
   magnetic, 220
Fiber optics, 201, 216
Filament, 205
Filter, 190
Fingerprints, 65
Flashlight, 209
Flow rate, 144
Fluids, 129
Fluid flow, 140, 144
Fluid pump, 145
Fluorescent light, 205
Focal length, 212
Focal point, 209
Foot-pound, 99
Force, 52, 86
   buoyant, 132
   centripetal, 101
   definition of, 52
   multiple, 57
   gauge, 53
   of gravity, 54
   sensor, 69
Four-stroke engine, 177
Fps, 15
Free fall acceleration, 42, 43
Freezing, 165
Frequency, 187
Friction, 62, 73, 140
Frictional force, 62
Fuel cell, 87
Fuels, 78, 173
Fulcrum, 108

G

Gas mileage, 78, 79, 173
Gasoline, 173
Gasoline engine, 177
Gas molecule, 173
Gear pump, 146
Gears, 97
Generation of electromagnetic waves,
      204
Generator electrical, 225
Gerbil, 234
Gravity
   acceleration of, 42, 55
   gravitational potential energy, 74,
      75
Gyroscope, 92

H

Hair dryer, 85
Hall probe, 232
Ham radio, 219